Simplified
TRIZ

New Problem-Solving Applications for Engineers and Manufacturing Professionals

Simplified
TRIZ

New Problem-Solving Applications for Engineers and Manufacturing Professionals

Kalevi Rantanen
and **Ellen Domb**

S^t_L

ST. LUCIE PRESS

A CRC Press Company
Boca Raton   London   New York   Washington, D.C.

Library of Congress Cataloging-in-Publication Data

Rantanen, Kalevi.
  Simplified TRIZ : new problem-solving applications for engineers & manufacturing professionals / Kalevi Rantanen, with Ellen Domb.
     p.   cm.
  Includes bibliographical references and index.
  ISBN 1-57444-323-2 (alk. paper)
     1. Engineering--Methodology. 2. Problem solving--Methodology. 3. Creative thinking. 4. Technological innovations. I. Domb, Ellen. II. Title.
  TA153 .R26 2002
  620'.0076—dc21   2002019233
                                                                                          CIP

Visit the CRC Press Web site at www.crcpress.com

No claim to original U.S. Government works
International Standard Book Number 1-57444-323-2
Library of Congress Card Number 2002019233
Printed in the United States of America 1 2 3 4 5 6 7 8 9 0
Printed on acid-free paper

PREFACE

People who need better tools for developing new systems, solving problems and selecting solutions include:

- Those doing research and development in technical and nontechnical fields
- Technology managers and other managers
- Everybody who solves problems

The book presents a new model for problem solving. The model is based on the theory of problem solving called TRIZ. TRIZ is showing up more and more frequently today in success stories of innovative solutions to problems in technology and in nontechnical fields. Many Fortune Global 500 companies such as Ford, Procter & Gamble and Mitsubishi have used TRIZ to develop better products more quickly. More and more small companies and individual inventors are using TRIZ. People in fields as diverse as marketing, education and management are using TRIZ methods to solve their problems. People who thought that creativity was a trait that some have and some lack have learned that through the TRIZ process everyone can be creative.

Many users of well-known improvement methods like the Theory of Constraints (TOC), Six Sigma, Quality Function Deployment, Taguchi method, DFM-A and others, have found that TRIZ is a valuable complement to the systems and methods they are using. TRIZ helps them use the other methods more effectively.

TRIZ comes from the Russian phrase *Teorija rezhenija izobretatelskih zadach*. The English translation is the Theory of Inventive Problem Solving. Why are people all over the world using a method with a Russian name?

TRIZ is an unusually global theory. It has a much wider multinational basis than most methodologies of management and creativity. The popularity of TRIZ in the U.S., Japan, Germany, UK, France, Korea, Israel and other leading industrial countries is not a surprise, as the use of innovation as a business strategy has been strong for at least the last two decades. (See the works of Hamel[1] and Porter.[2,3])

TRIZ has its origins in the former USSR, where it was founded by Genrich Saulovich Altshuller (1926–1998). He lived most of his lifetime in Baku, except the years 1950–54, when he was confined in prisons and camps and his last years in Petrozavodsk, northern Russia. In research on patents that started at the end of World War II, Altshuller found that a variety of different engineering systems and technologies had common patterns of evolution. The inventor (or any problem solver) can learn these patterns of evolution, use them to develop new technology consciously and systematically and avoid many fruitless trials and errors. His ideas about improving the way work was done didn't please Stalin's administration and he was arrested in 1950. In 1954, after Stalin's death, he was "rehabilitated" and returned to Baku. He continued to work on the new theory with a wide circle of colleagues. He conducted seminars and courses, mainly for engineers and wrote books and articles. Over the next 30 years, the TRIZ methods developed. TRIZ was taught in the universities, military academies, elementary schools, high schools and in a network of independent TRIZ schools.

Very little was known of this work outside the USSR for more than 40 years. In the early 1990s, it became possible for Soviet experts to travel abroad and for some to emigrate. A TRIZ boom rose in the U.S., Japan and many other countries, in part because of the availability of software to support the use of TRIZ in English. Many new users, researchers and service providers became involved. TRIZ became global. "Russian stuff" was mixed with customer-oriented approaches in the West. The result has been an extremely fruitful combination of Eastern and Western traditions.

The word "TRIZ" is the name of the theory, not a trademark. As TRIZ has proliferated, very different and sometimes contradictory things have been presented under the label TRIZ. This book will be different.

If we compare TRIZ with the automobile, this book is for drivers. You don't need to become an automotive engineer or mechanic to drive the car. You don't need to become an expert in the methodology to use TRIZ. If people tell you that TRIZ is complex, don't get worried and don't believe them. This kind of criticism probably reflects experience with the old TRIZ. At the birth of the automotive industry, drivers needed to be their own mechanics. At the birth of the computer era, only programmers could use computers. TRIZ has followed this pattern — the first TRIZ users were

TRIZ experts. The situation is different today. Now everyone can learn TRIZ and use it effectively very quickly.

This book presents modern, "international" TRIZ. It is not a translation or review of older Russian books. The core concepts, which have been selected from the work of Altshuller and his colleagues, are as follows:

- Contradiction
- Resources
- Ideality
- Patterns of evolution
- Innovative principles

These concepts have passed the difficult test of the market. The authors have tested them with their students and in their consulting work. They can be understood and used rapidly by beginners and they are valuable to experts as well.

On the other hand, much material traditionally included in TRIZ texts has been left out. Long step-by-step guides or "algorithms" are avoided. Simple models have replaced some outdated and unnecessarily complex ones. Many good and interesting things traditionally included in TRIZ books have been left out because they are not useful to people who want to start using TRIZ quickly. If you need information not included in this book, you can find it easily by consulting the references provided in each chapter, especially from The TRIZ Journal,[4] which is a free online resource.

This book is a practical guide. It is a "how-to" book. It shows you and tells you how to solve problems creatively and — this may be even more important — shows you how to find problems and foresee the evolution of both the problems and the solutions. The book contains many exercises, worksheets and tables. You can download blank copies from http://www.triz-journal.com/SimplifiedTRIZ/.

At the same time, this is a strongly scientific work. The basic concepts and the models connecting them are emphasized. The word "scientific" means that the readers themselves should test and refine the generic model. We advise each reader to:

- Not accept TRIZ only because it is in fashion or because well-known corporations support it.
- Accept and embrace TRIZ because it works for you and for your problems.

The results of behavioral sciences, especially findings of activity theory and cognitive psychology, have been used to develop this book. Its structure is designed to guide the reader through a successful learning and implementation process. Studying the book will take you through six stages:

1. Motivation (why do I need new tools?)
2. Orientation (forming a general vision or mental model)
3. Internalization (enrichment of the mental model or getting new knowledge)
4. Application of the model to your own problems
5. Evaluation (testing and refining the model against your own experience)
6. Implementation (modifying the general process to work in your environment)

After studying this book, you will be eager to tell other people in your organization about TRIZ. To convince others, you should first convince yourself. For that, you need your own examples and cases. If you complete the exercises throughout the book, you will have a set of examples based on your own work. You can use each chapter separately and learn each tool or you can work with the entire system. You will find that each tool is independently useful, but when used together, the system is even more helpful. Chapter 12 provides a road map for how to introduce TRIZ into your organization. Doing the exercises will help you help others to appreciate the need for TRIZ, which is the first step toward implementation.

The authors first met as authors and editors of The TRIZ Journal, a free online monthly resource for the TRIZ community. We invite all our readers to sample what others have done around the world by reading the journal and by contributing their experiences to it throughout all six stages of learning.

REFERENCES

1. Hamel, G., *Leading the Revolution*, Harvard Business School Press, Boston, 2000
2. Porter, M.E., *On Competition*, Harvard Business School Press, Boston, 1998
3. Porter, M.E., *Strategy and the Internet*, Harvard Business Review, March 2001
4. *The TRIZ Journal* http://www.triz-journal.com

ACKNOWLEDGMENTS

Thanks are due to Pekka Koivukunnas from Metso Paper Corporation, who has offered many valuable user comments during the preparation of this book, as has Veli-Pekka Lifländer from Espoo-Vantaa Institute of Technology. Timo Saraneva, a friend and colleague, has helped to develop the appropriate model and Ralph Czerepinski, Tom Kling and Gregg Motter of the Dow Chemical Company made a number of contributions to our knowledge of how to teach TRIZ to people with a variety of different interests. And finally, we acknowledge the contributions of Phil Samuel and Dan Laux and their colleagues at the Six Sigma Academy, who are pioneering the application of TRIZ in the Six Sigma process.

We would also like to thank our spouses, Galina Rantanen and Bill Domb, for their patience and support and many creative suggestions throughout the process of developing *Simplified TRIZ*.

AUTHORS

Kalevi Rantanen is a Finnish TRIZ expert who successfully combines many different experiences and areas of knowledge in his work. In the 1970s, he worked in youth organizations, mainly on the problems of education and training. In the early 1980s, he studied in the former USSR, earned his M.Sc. in mechanical engineering and discovered for himself an unexpected, very exciting new world: TRIZ. He has worked in industry since 1985 and, since 1991, as an independent entrepreneur. He concentrates mainly on TRIZ training as the bridge between the leading theory of creativity and business objectives.

Ellen Domb is the president of The PQR Group, a U.S. consulting firm specializing in helping organizations maximize customer satisfaction, productivity and profits through strategic management of quality and technology. Formerly a director of the Aerojet Electronic Systems Division with specific responsibility for Total Quality Management implementation, she is a founding board member and Judge for the California Council on Quality and Service. She is a charter member of the Quality Function Deployment Institute, co-founder of The TRIZ Institute, editor of The

TRIZ Journal (www.triz-journal.com) and chair of the first English language International TRIZ Symposium.

CONTENTS

1

WHY DO PEOPLE SEEK NEW WAYS TO SOLVE PROBLEMS?

INTRODUCTION

In this book, we will study how to generate and select good solutions to problems using TRIZ, a new theory of problem solving. The term "TRIZ" comes from the Russian phrase *teorija rezhenija izobretatelskih zadach* which means the "theory of inventive problem solving."

But why do we need a new theory? Without a theory, people generate ideas by guesswork and then select the ones they like or those they think other people will like. With TRIZ, you will be able to generate better ideas faster and you will have a basis for selecting the best ideas, the ideas that will solve your problem effectively and form a basis for further improvements. In this chapter we show that good ideas have frequently been rejected when first proposed. Much money and time and human effort is lost when good ideas are rejected.

We show that people cannot select ideas properly and cannot produce better ideas effectively if they are unaware of the common features of good solutions: resolving contradictions and making use of idle resources. We show that common, traditional approaches to problem solving have often turned out to be dead ends. We need TRIZ, the theory based on the features of great inventions and the patterns of the evolution of systems, rather than approaches with no theory, based on the psychology of people.

1.1 WHY ARE GOOD IDEAS REJECTED?

People create new technologies and make creative use of existing technologies, generating lots of new ideas. But how can we know which idea is good and which is not? History shows that companies and society as a whole have frequently rejected good ideas and invested money in ineffective ideas. Consider some examples:

1. Alexander Fleming observed in 1928 that a mold culture produced something that was poisonous to many hazardous bacteria and not to humans. He named the new substance penicillin and published his results in 1929 in a well-known professional journal. In 1938, Ernst Chain read Fleming´s article and became interested in penicillin. In 1939, he got a $5000 grant from the Rockefeller Foundation for the development of the new drug. It was the beginning of the penicillin industry. Why did scientists and investors ignore penicillin for 10 years? How was it possible that medicine and the drug business so long preferred older, often hopelessly ineffective drugs and therapies to penicillin?

2. T-DRILL is a manufacturer of tube and pipe fabrication machines. The basic idea is simple: a collar is formed as part of the tube, from the material of the tube, replacing a conventional T-fitting. Only one joint is needed instead of three. The benefits of T-joints without T-fittings in many applications are now indisputable. It took about 30 years to get this easy-to-understand idea accepted. Why?

3. In 1948, Dick and Mac McDonald opened a fast-food restaurant where the customers themselves performed the function of waiters. In 1954, Ray Kroc took a look at the McDonalds' stand. He saw that never had so many people been served so quickly. He understood the fast-food concept immediately. But why did it take 6 years for an entrepreneur to understand the concept? Why did the great majority of restaurateurs continue to offer poor service at higher prices?

4. Molok is the trade name for a new bin for garbage and recyclables. The principle is simple. Molok is a vertical container, partly hidden underground — only 40% of the container is visible. The weight of the waste is used to compress the waste; you could say that gravity does the work. Because the new bin is partly underground, there is no odor. The container is lined with a big bag that can be removed and transported without complex specialized machinery. Why was this simple innovation not introduced until the 1990s? And why has it met considerable resistance?

5. Flash smelting technology for copper, introduced by Outokumpu in the 1940s, is one of the most successful innovations in metallurgy in the 20th century. Here, too, the general principle is simple. Sulfur contained in the ore enhances the fuel efficiency for smelting. The need for energy from outside is reduced

Table 1.1   Examples of Good Solutions

Year created	Year implemented	Idea
Note: If you prefer not to write in the book or if you want extra copies of the tables, they can be downloaded from http://www.triz-journal.com/Simplified TRIZ/.		

drastically. Outokumpu's competitors also knew very well that if the ore contains sulfur, there is free energy available for smelting. Why did they ignore what they knew?

Every industry has examples like these. What are some examples from your industry, of good ideas that were ignored when first introduced? Use the exercise, Table 1.1, to collect examples of good solutions. Copies of workbooks and tables are available from www.triz-jounal.com/simplified TRIZ/.

> *Why does it take years (or decades) for so many excellent solutions to be used, even though they are urgently needed and the technology is available?*

This question has been asked many times in TRIZ classes, presentations and discussions. The audiences usually offer some form of the following answers:

■ The inventor is seldom a good salesperson.
■ Lack of support from management.
■ Poor presentation of the idea.
■ Prejudices or the popular buzz-words "paradigm paralysis."
■ NIH (the not-invented-here syndrome).

At first glance, the answers are self evident. Closer examination reveals, however, that these answers do not help much. Imagine that all inventors become good salespeople, have the support of management and have excellent presentation skills and materials. But how can the inventor, product developer or management know what idea is worth promoting? Companies have often used excellent sales skills to support outdated products. Richard Foster's book *Innovation*[1] gives many good historical examples. National Cash Register continued to advertise electromechanical cash registers in the 1970s when the development of electronics had already made them obsolete. The producers of cross-ply tires for cars were very customer-oriented. This didn't help when the superior belt tire captured nearly the whole market in a short time. The list of examples can be easily continued: sailing ships vs. steamships, vacuum tubes vs. the transistor, conventional bike vs. mountain bike and so on. The crucial point here is that content matters. It is trivial to say that inventors should be able to get their new ideas accepted, but how can they know what idea is really new and better than the old technology?

What about prejudices? Would the result be better if experts and managers were less prejudiced and more creative? It is true that the inventor should be open to new ideas and criticism. Every idea needs a champion who can fight stubbornly against resistance and indifference. But how can inventors know when to accept ideas or criticism and when to reject them? Again, the content of the idea is important. One must select a good solution from many ideas, some good and some bad. But how can we select the best solution? Is it best for our customer? Best for our business? Is it the most interesting technology?

We propose a simple reason for the rejection of good ideas: People reject good solutions and invest in bad ones because they don't know the difference between them.

Looking at cases of lost opportunities and great losses for business and society, people can take one of two positions:

1. With hindsight, it is easy to see that often very good ideas are rejected and resources lost to bad ones. But obviously it is not possible to know whether the idea is good or bad when it is first proposed.
2. Because the same patterns are repeated over and over again, we can learn from the past. The patterns of evolution can be discovered and used to get better solutions today.

There is growing evidence supporting position 2, using TRIZ to provide the general theory of the evolution of good ideas. For example, the statement from a participant of a TRIZ class conducted by one of the present authors (names are removed, the rest is cited verbatim):

"A sad but true testimony to the power of TRIZ. In one of our TRIZ sessions that you conducted at XXX, we identified the use of water, transformed into steam, as a method for foaming an adhesive. This idea, though considered valid, was never acted on. A patent was recently issued to one of our competitors for a process of foaming an adhesive with water vapor."

1.2 COMMON FEATURES OF GOOD SOLUTIONS

Good solutions have several common features. The good idea does the following:

1. Resolves contradictions
2. Increases the "ideality" of the system
3. Uses idle, easily available resources

In addition to their everyday meanings, these words have specific technical meanings in TRIZ. By working with the TRIZ concepts, you will learn to apply them to your problems, to develop good solutions and to select the best solutions from all that are proposed.

1.2.1 Three Basic Concepts for Reaching the Best Solution

1. A good solution resolves the contradiction that is the cause of the problem. There are two kinds of contradictions:
   Tradeoff contradiction means that if something good happens, something bad happens, too.
   Inherent contradiction means that I want that one thing that has two opposite properties.
2. The "ideality" of a system is the measure of how close it is to the perfect system. The perfect system (called the "ideal final result" in TRIZ) has all the benefits the customer wants, at no cost, with no harmful effects. So a system increases ideality when it gives you more of what you want or less of what you don't want, does it at lower cost and usually with less complexity.
3. Unseen idle resources of the system are used to reach these seemingly incompatible goals. These resources include energy, materials, objects, information or things that can be made easily from the resources that are in the system or nearby.

All five examples of resistance to new technology illustrate clearly these concepts of overcoming contradictions, increasing ideality and using resources:

1. Penicillin resolved a typical contradiction of drugs: substances that can kill microbes destroy healthy tissues, too. Mold, present every-

where, was used for resolving contradiction. Increasing ideality: many important diseases, earlier considered totally hopeless, were easily cured by penicillin.

2. Collar made as part of the tube: there is no separate T-fitting. The tube is used as a resource. Ideality increases: one joint is less complicated, requires less material and uses less labor than three.

3. Fast-food restaurant: there are no waiters, but, at the same time, all customers have their own waiter, i.e., serve themselves. Resource: a customer. Increasing ideality: better service more quickly.

4. A bin for garbage should be big and at the same time small. A partly underground container is big (in available volume) and small (the part you see). The space under the bin is an easily available resource.

5. Flash smelting resolves a contradiction: much energy is needed and energy should not be used at all. Sulfur in the ore is an easily available energy resource.

These three concepts are repeated again and again. Contradictions are solved. Idle resources are used. Solving contradictions by using resources makes the system more ideal. We can describe the movement from the problem to the solution by a simple diagram (Figure 1.1).

We see that creative activity in research and development (R&D), manufacturing, marketing, management and other areas needs reorganization. A simple scheme (Figure 1.2) shows the desired changes.

1.3 A NEW APPROACH TO PROBLEM SOLVING IS NEEDED

There have been, of course, many attempts to make creative work more effective and to replace the trial-and-error method. These approaches have different names. However, they can be easily divided to two groups:

1. The first group can be called the rationalized or "hard" model. McGregor's Theory X describes this model well.[2] Research and development centers are established. The work is controlled by budgets and time limits. The silent assumption is that people need to be controlled and directed rather tightly. This kind of management helps to get minor improvements, but seldom gives great, qualitatively new ideas.

2. Many attempts have been undertaken to overcome the weaknesses of the rationalized model. Many "creative" techniques have been offered. There are few substantive differences between

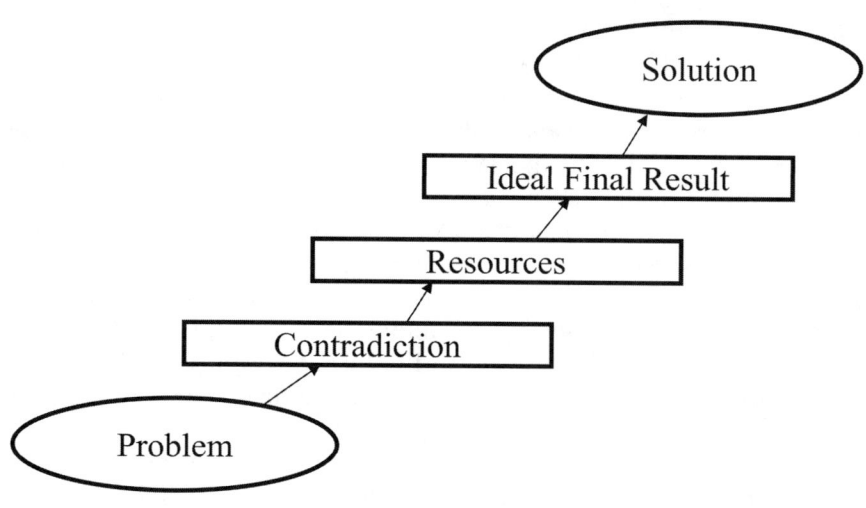

Figure 1.1 Features of good solutions. Contradictions are solved. Idle resources are used. Solving contradictions by using resources makes the system more ideal.

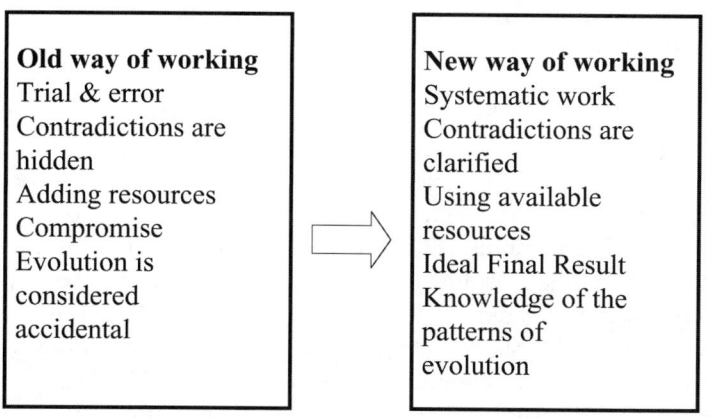

Figure 1.2 The reorganization of creative activity. The transition from the old way of working to the new.

them. Together they can be called a humanized or "soft" model, which fits McGregor's Theory Y. In this model, people naturally have imagination and creativity. External control is not the only means for getting good results. So criticism and control are minimized or prohibited. Fantasy, feeling, play, intuition and pleasure are encouraged.

The humanized model is often attractive in the beginning. Many ideas are generated. Soon, however, most proposals turn out to be the repetition of old inventions. Sometimes really good ideas are developed, but they are not recognized due to the lack of evaluation criteria (see examples in the beginning of this section).

Christopher Freeman[3] has characterized *demand-pull* theory and *science-push* theory as two poles in the debate on the determinants of innovation. Rationalized activity and humanized activity are considered two opposite means to improve the traditional craft activity, as described by Engeström.[4] Theory X and demand-pull theory can be loosely related to the rationalized model, Theory Y and science-push theory to the humanized model.

Disappointment in the "soft" approach causes organizations to return to the "hard" model. Then, after some time, traditional management is criticized for the lack of creativity and "free" idea generation comes into fashion again. And so on. Inventive methodologies in industry seem to oscillate perpetually between hard and soft models. Both ways are blind alleys. Both are unsatisfactory.

There are, of course, many good strong tools for the development of systems, but they all seem to be missing specific techniques for problem solving. Such methods include:

- Quality Function Deployment (QFD) identifies the voice of the customer and helps the organization understand where creative ideas are needed. But it has no tools to create new concepts to meet the customers' often-contradictory requirements.
- Theory of Constraints (TOC) helps to define conflicts and to identify where a conflict resolution is required, but does not have tools and techniques for generating the ideas that will resolve the conflict.
- In 10 years, Six Sigma has gone from "just another quality system" to a corporate management system that gets Wall Street's attention for its ability to mobilize organizations. Six Sigma integrates many methods of problem identification and analysis, but it took until 2001 for some organizations that teach Six Sigma to begin incorporating TRIZ to get good solutions to the problems that were identified.

The last chapters of this book will deal with incorporating TRIZ into these methods, so that organizations can combine the power of

TRIZ for solving problems with the power of any of the methods of finding problems.

The new approach, neither hard nor soft, but incorporating the benefits of both views, has become necessary. It is not a mechanical sum of traditional ways to think. It is the TRIZ system of understanding the problem, modeling the contradictions and removing them by using resources and improving the ideality of the system, not relying on intuition. It relies on knowledge of the system being improved and on knowledge of the systematic method for improvement.

TRIZ is based on more than 50 years of research, but it is new to most of the industrial world. Increasing consciousness of the weakness of traditional approaches has increased interest in TRIZ. TRIZ does *not* ask, What is the difference between creative and uncreative people or organizations? TRIZ asks: What is the difference between good and bad *ideas*, solutions and products? TRIZ seeks the sources of creativity in the objects or systems to be improved in the outer world, not in the psychology of the people or the organization doing the work.

A simple comparison illustrates the approaches. Runners can increase their speed using physical and mental exercises. A coach can manage runners using the hard or soft way: compel them to do structured exercises or give them freedom to run however they want. Both methods have been used to increase speed and certain methods work better with certain runners. The speed can be increased, but not very much and not very quickly. A different way is to say that the goal is to go fast and to provide the runners with vehicles: bicycles, cars, planes and boats. Now the main point is not differences between people, but differences between tools and everyone can go fast. TRIZ offers vehicles for moving to better ideas, solutions and innovations. Knowledge of the features of good solutions is the "vehicle" that can be used to generate better solutions. The point is not to learn more about the psychology of people in order to increase creativity, but, using TRIZ, to develop creative ideas, no matter what kind of intuitive skills each has.

All four approaches we have considered are presented in a simple schematic drawing (Figure 1.3). These models describe a generic framework. If you model your own experience, it will give this framework more meaning for you. The exercises in Table 1.2 will help you do this. Start by filling in a description of the methods that your organization has used for problem solving and for stimulating innovation. If you have gone back and forth from one method to another, draw arrows on the table to show the path. If you have tried TRIZ or a method related to TRIZ, list it under Scientifically Managed Problem Solving and use arrows to show the path from other methods to TRIZ.

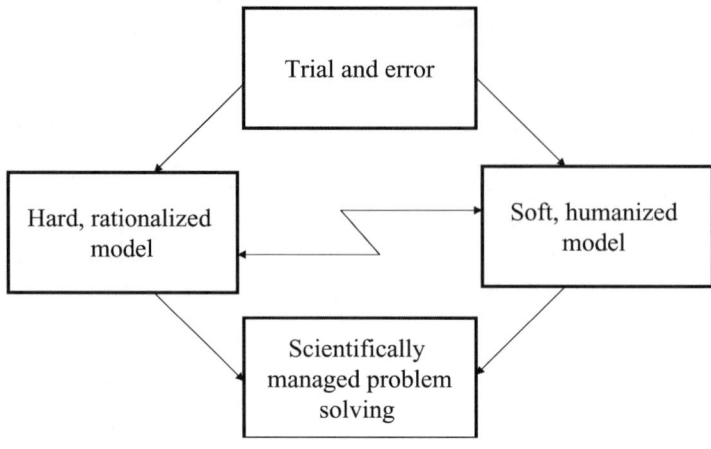

Figure 1.3 Models of creative work. Trial-and-error method is the oldest way of seeking ideas. "Hard" and "soft" models are two popular attempts to make work more effective. Scientifically managed problem solving combines the strengths of both approaches with the theory of TRIZ.

Table 1.2 Methods Used by Your Organization for Problem Solving and Stimulating Innovation	
Trial and Error Methods	
Hard, rationalized model	*Soft, humanized model*
Scientifically managed problem solving	

1.4 SUMMARY

- To generate and select ideas for good solutions to problems, one should know the difference between strong and weak ideas. To promote ideas, you should know which solutions are worthy of being promoted.
- Traditional methods of problem solving did not have criteria for selecting good ideas.
- A strong solution resolves a contradiction, makes use of idle resources and increases the ideality of the system. TRIZ is the theory that provides the basis for this model of successful problem solving.

The next chapter will give the big picture of the new model for problem solving.

REFERENCES

1. Foster, R.N., *Innovation: The Attacker's Advantage*, Summit Books, New York, 1986.
2. McGregor, D., *The Human Side of Enterprise*, McGraw-Hill, New York, 1960.
3. Freeman, C., The Determinants of innovation, *Futures*, June 1979, 206.
4. Engeström, Y., *Learning by Expanding* orienta-Konsultit, Helsinki, 1987, 284.

2

CONSTRUCTING THE NEW MODEL FOR PROBLEM SOLVING: MOVING FROM THE PROBLEM TO THE IDEAL FINAL RESULT

INTRODUCTION

In the first chapter, we showed that a new approach to problem solving is needed and briefly outlined the basic features of TRIZ. In this chapter, we construct a model for problem solving. The model is like a general map that shows, by words and pictures, how to use the most important TRIZ features in problem solving. Details will be studied in later chapters. This short chapter is very important. The model presented will guide you through the details and help keep you on track as you study and use TRIZ.

We will construct the model in five steps. First, we describe the concept of contradiction. Second, mapping resources is added to the model. Third, the concept of the ideal final result is formulated. These steps form the inner shell of the model. The fourth and fifth parts of the model are the patterns of evolution and innovative principles.

This model for problem solving is based on the theory of TRIZ, on customer feedback from people who have used it and on the knowledge of the styles of human thinking and problem solving activity.

2.1 CONTRADICTION

2.1.1 Difficult Problems Contain Contradictions

One of the early insights of the TRIZ researchers was that solving a problem meant removing a contradiction. If we compare the TRIZ problem-solving methodology to a tree, the concept of contradiction can be compared to the seed, from which we can grow the whole tree. If we would like to express the idea of TRIZ by a single word, that word would be "contradiction."

A contradiction is a conflict in the system. A system consists of two components: tool (T) and object (O). The edge of the ax blade, for example, is a tool that splits the object, a chunk of wood. Splitting power is a good feature that is connected with harmful properties, like the clumsiness of the tool. If you make the blade heavier, the ax can strike a more effective blow, but it becomes more awkward to handle.

We meet contradictions everywhere. For example, a company wants to improve business by improving customer service and decides to get better service by increasing staff training. Training is the T that is used to improve a certain O: the professional quality of employees. If employees get extensive and thorough training, service surely can be improved, but the time loss might be intolerable. We get a contradiction: the better the service, the more training time is needed.

Think of a seesaw (Figure 2.1). When one end of the plank goes up, the other goes down. You cannot get both ends to go up at the same time.

In this case, the connection of features (up-down) is not a problem — having one end go up when the other goes down is a natural property of a seesaw. If you don't want the experience of going up and down, you will need to choose a different toy or make some major changes to the seesaw system.

The first three concepts can be illustrated by a simple figure (Figure 2.2) containing a tool, an object and contradiction.

We will use a diagram like this for each problem. A flash-like arrow between a tool and an object indicates a contradiction. Many visual, mathematical and physical models are available for the design, use and

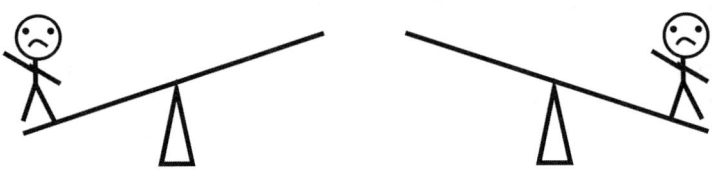

Figure 2.1   The seesaw analogy of the problem situation.

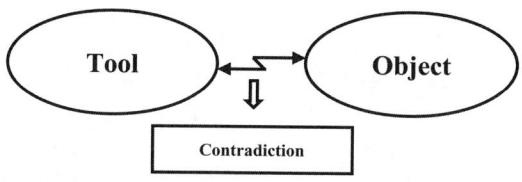

Figure 2.2   Tool, object and contradiction. Contradictions between tools and objects are the moving force of evolution.

maintenance of systems. This simple model is easy to use in deciding whether you need TRIZ to solve the problem. Drawing the diagram makes you think about the problem and decide whether you have a contradiction. You can solve many kinds of problems using TRIZ, but the theory is most powerful and gives the most value added when used to solve non-routine (that's the "inventive" in "theory of inventive problem solving") problems containing contradictions.

2.2 RESOURCES

Sometimes the clear formulation of the contradiction suggests a possible answer to the problem. Usually, however, additional information is needed. Resource analysis helps you find ways to resolve the contradiction.

Resources are things, information, energy or properties of the materials that are already in or near the environment of the problem. If they can be used directly or modified to make them useful, the problem will appear to have solved itself. Think of the resources as the reserves — they are invisible at first, because we are accustomed to not seeing them when we look at the problem situation, but we can mobilize these reserves to solve the problem. That's why we add the block "Resources" to our model (see Figure 2.3).

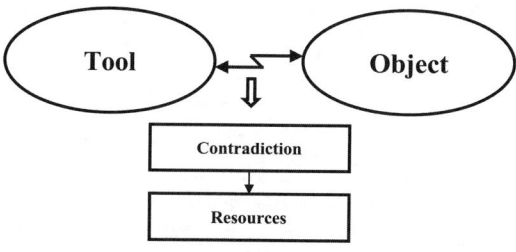

Figure 2.3   Resources are information, energy, properties, etc., available for solving contradictions. They are often invisible at first, because we are accustomed to not seeing them when we look at the problem situation.

The system should be changed so that the needed improvement seems to appear from nowhere. For example, in Chapter 1 we described a garbage bin that was partly hidden underground. The needed change was achieved using the space beneath the bin.

In the ax example mentioned earlier, we need to change the system so that splitting power is improved but the ax does not get more difficult to use. The resources in the ax and chunk of wood system are the blade, its edge, its form, material and other properties; the handle and its properties; the chunk of wood and its properties; the surrounding air, etc. Depending on how the system is defined, we can also include the person whose arm provides the energy and whose hands provide the transmission of energy to the ax.

In the training example, the curriculum, its structure and properties, skills of teachers, textbooks, the motivation and existing knowledge and skills of the students are all resources, as well as the culture of the company and the physical environment of the training.

2.3 THE IDEAL FINAL RESULT

Using resources, one can remove the contradiction and get the ideal final result. This is the final concept that will be added to the model (Figure 2.4).

Imagine that both ends of the plank forming the seesaw go up by virtue of resources. If the child's resources included some rope, a tree and a strong parent, both ends of the plank could go up at the same time (Figure 2.5).

The blow struck by the ax should be made stronger but, at the same time, the tool should remain easy to use. To split the chunk of wood, the

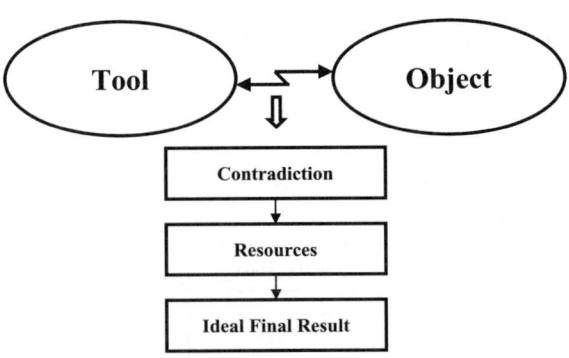

Figure 2.4 The ideal final result is the solution that resolves the contradiction without compromise. Resources are used to go from the contradiction to as a perfect solution as possible.

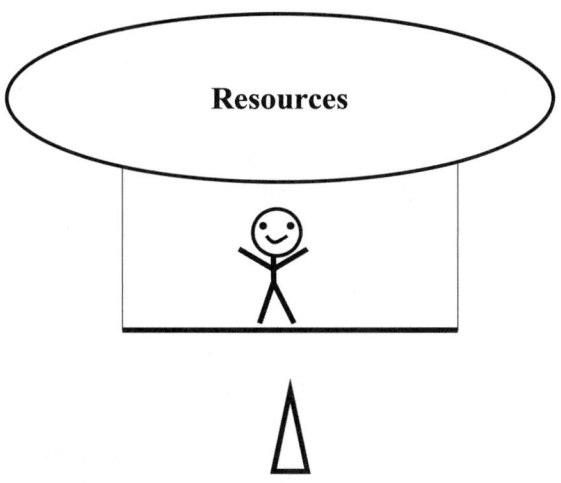

Figure 2.5 The seesaw analogy of the ideal final result. Both ends of the plank go up.

ax should be heavy but, at the same time, for ease of handling, the ax should be light.

The ideal final result can now be described as follows: something changes the ax in some way so that it is both heavy and light, for the double purpose of making the blow stronger without decreasing the ease of use. The ax marketed by Fiskars solves this problem by using a hollow handle. Air is used as the resource that solves the problem. The hollow handle gives an unexpected new quality. The center of gravity moves nearer to the blade. The blow is more powerful, which could be thought of as "heavy," although the tool is lighter.

In the ideal training, results are improved but the length of time is not increased. Changing the training so that part of it is done on the job, not in the classroom, frequently results in better learning as well as shorter classes, because the students can see the results immediately and are in a better position to apply them. This improves both motivation and feedback.

"Ideality" is the measure of how close the system is to the ideal final result. If the useful feature improves, the ideality improves. If the harmful feature lessens, the ideality also improves. To solve the problem, look for resources already in the system that can help the useful feature get better or the harmful feature be reduced — or vanish entirely.

2.4 PATTERNS OF EVOLUTION

Formulate the contradiction, map resources and define the ideal final result. Is this enough? Sometimes yes. Often, however, something more is needed to move from the ideal final result to the technical solution of the problem. We need methods to follow to resolve contradictions, to use resources and to make the system more ideal. One method is the use of the patterns of evolution of systems. We will speak in this book of features, patterns and laws. Features are, as in everyday language, any properties of systems: size, weight, speed, flexibility, color, etc. Patterns of evolution are important regularities in the development, for example, transition from the macro- to the microlevel or the division of the system to smaller parts. Patterns are actually laws, but they are soft, not rigorous mathematical formulas as in physics. That is why we usually refer to them as patterns.

Studies of the history of innovation have shown that many improvements follow similar patterns. We just named the transition to microlevel or division of the system into parts. Let's look around. In your computer and printer, electrons and other microscopic parts have replaced the cumbersome details of old typewriters and calculators. In the kitchen, you have a microwave oven. The food is heated by electromagnetic vibration of its water molecules, not by radiant heating from a heavy metal plate. You may work in an organization that has divided itself into many relatively independent teams. If you like outdoor activities, you can wear clothes made from microfibers. In your household work, you may use microfiber cloths for cleaning. The bed in your bedroom is also making the transition to microlevel. Water beds, air beds or mattresses composed of small cells of some kind in place of steel innersprings are attacking traditional beds. Everything seems to get divided into smaller parts.

Different parts of a system may change at different rates and may follow different patterns. The evolution is uneven. We spoke of the transition to microlevel. There is the transition to macrolevel (also called the supersystem), too. The parts of the system become more interactive with each other. The system is expanded and convoluted. It is improved by adding more and more features, then combining all the features into a new, simpler system that has all the benefits without all the complexity.

We can use these patterns to find hints about how any situation can be improved and to obtain suggestions about how the system could be changed to become ideal. Chapter 9 is a detailed presentation of how to use the patterns of evolution.

In the ax example, one can think of the transition to the microlevel. The ax can be segmented or divided into parts. If you continue segmenting it into smaller and smaller parts, eventually you get particles, then molecules. An ax made of molecules? A gaseous ax? Is this crazy? The hollow ax, containing air in the handle, is partially gaseous. Or, a stream of particles, like a sandblaster, can be a very effective cutting tool, where each particle could be considered a microax.

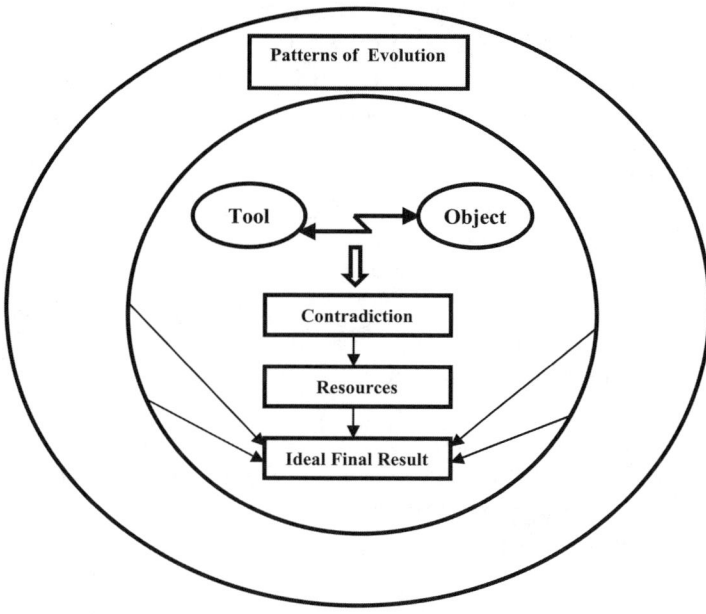

Figure 2.6 The patterns of evolution have multiple uses: they support the tools for problem solving, help to get solutions directly and can be used for the prediction of the future features of technology.

Some important patterns of evolution can be applied to business systems. Training is a system that evolves unevenly. A training program can be segmented (transition to microlevel) and integrated to larger systems (transition to macrolevel). Programs tend to expand, then inevitably be compressed so that efficiency is maintained or increased. The simplified model of TRIZ that we are constructing here is an example of trimming a complex subject to make it easy for people to get started using it.

We add the patterns of evolution to the pictorial model. They can often be used directly to develop good solutions to problems, as well as to predict the future evolution of the system. Figure 2.6 illustrates multiple uses of the patterns of evolution. To keep the model simple, arrows are drawn only from the patterns of evolution to the ideal final result. This connection is most important for problem solving because the more ideal system is what we need.

2.5 INNOVATIVE PRINCIPLES ACROSS INDUSTRIES

We now have four important concepts in the model: contradiction, resources, the ideal final result and the patterns of evolution. But these are not always enough. The evolution pattern of the system may suggest a rather vague idea of solution and we may need much more specific help. Innovative principles are tools that tell what the patterns mean and help us interpret the patterns for any particular problem. There are 40 principles, which will be studied in detail in Chapter 10. A few examples show how they are used.

For the example of the improvement of the ax, it might not be obvious at first that the replacement of a solid body with a hollow one follows the pattern of transition to a microlevel. Principle 29 gives a more concrete hint: "Pneumatics and hydraulics. Use gas and liquid as parts of an object ...." Now one needs to remember only that air is also gas, to see that a part of the ax can be made from air. The idea of the hollow handle is generated nearly automatically.

The 40 principles are based on the same study of patents and technology that developed the patterns of evolution. In the larger sense, this is a study of how people solve problems, not just a study of technology. That is why the same principles can frequently be applied to problems in management, marketing, training and other fields, even though there are no corresponding selected knowledge bases for creative problem solving. For example, Principle 18: Mechanical Vibration, suggests using an object's resonant frequency. It may mean resonance or synchronization in mechanical, electromagnetic or acoustical systems, but in training, it can also be interpreted as improving the coordination of the curriculum, textbooks and teaching with the learning style of the students and the culture of the company.

Let's add principles to our model (see Figure 2.7). The arrow from the principles to the ideal final result shows that these are shortcuts that sometimes allow us to bypass the analysis of contradictions and resources. Using the whole model is more effective than using the parts separately. Innovative principles are studied in detail in Chapter 10.

2.6 OTHER CONCEPTS AND TOOLS

Figure 2.7 is the model that will be used throughout this book. It is easy to remember, easy to use and will give you the power of TRIZ very quickly. Traditional TRIZ has many more tools that you may want to explore after mastering the tools and concepts of the model in this book. Figure 2.8 shows an enhanced model that you can use if you study any of the other TRIZ systems later.

The model also shows four other tools: ARIZ, standards, effects and TRIZ-based software. ARIZ is a long step-by-step guide for the analysis and resolution of contradictions (Algorithm for Inventive Problem Solving). Standard solutions is a list of the ways of transforming the system, based

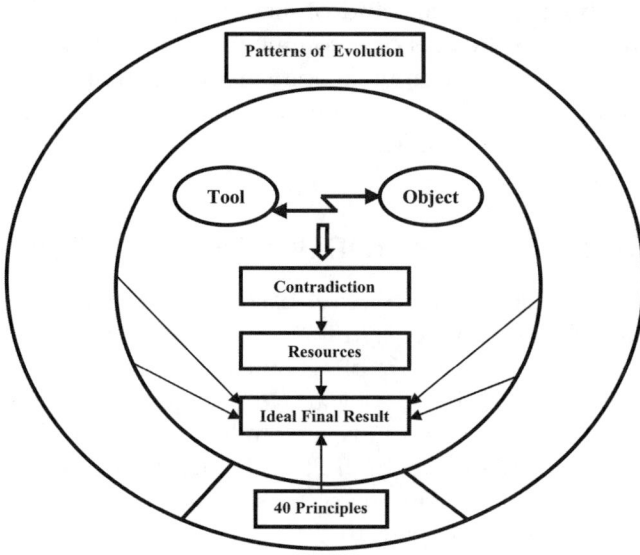

Figure 2.7 Forty innovative principles give cues for finding ideas. They can be used both as independent tools and to support other methods.

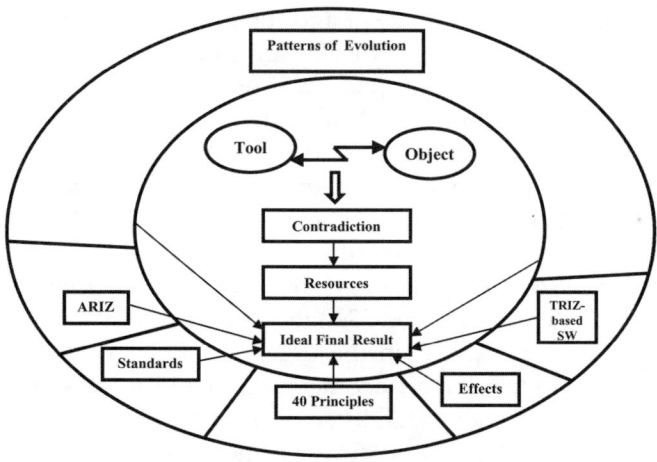

Figure 2.8 Many other tools can be added to the model.

on Altshuller's 1985 publication of a list called 76 standards. (But one of them has 10 parts.) Effects is a technical database of physical, chemical, mechanical, biological, geometrical and other technical phenomena that can be used for inventive problem solving. Various forms of the database appear in textbooks, software and on-line resources. Standard solutions, principles and effects are lists of recommendations and examples that can be easily processed by computer. Software containing these and other TRIZ-related knowledge bases has been available since the early 1990s.[1–3] Many other tools and techniques are available from other books and from on-line resources.[4]

2.7 WHY INTRODUCE THIS MODEL?

The model presented in this book is based on modern TRIZ, on recent achievements of innovative design in industry and on the latest research. The research on the evolution of technical systems and other systems guided the creation of the model and the organization of tools in the model. At the same time, we have carefully kept intact some important old tools that have proven to be fruitful, handy and robust during many years of application. Most famous of these classical tools is the list of 40 principles presented in Chapter 10.

The second reason that we have begun with the general concepts described by the model is the feedback from users. Users we have met in our training classes prefer general concepts to long procedures. They have told us that they prefer to study concepts first, then short step-by-step guides and checklists, then apply the concepts to their own problems. They like the contradiction concept and the ideality concept as new lenses for seeing reality. They usually dislike long instructions — it seems that an instruction longer than one page will never be used.

The third reason is the results from behavioral sciences, from the study of human activity. Activity theory and cognitive psychology have shown that individuals and teams need general organizing models to solve problems effectively. Work researchers use a model of human activity containing subject, tools, object (there is considerable overlap with the definition of a system in TRIZ) and also community, rules and division of labor, see Engeström.[5] Peter Senge and his team speak of "mental models" and "shared visions" in the organization.

All kinds of models have become fashionable in recent years. Many different words are used: model, internal model, mental model, paradigm, vision, schema and others. It is, however, not enough to say that internal models are necessary. The model should adequately describe the essential features of the object. In his article on the history of the transistor, Shockley describes the foyer of the main entrance to Bell Laboratories, where the following statement credited to Alexander Graham Bell is posted: "Leave the beaten track occasionally and dive into the woods. You will be certain to find something that you have never seen before."[6]

The statement encourages traditional problem solving through the trial-and-error method. The mental model of TRIZ can be compared to a map and a compass. One leaves the beaten track, but not with empty hands. The model is sometimes called an orienting basis by direct analogy with orientation in the forest using maps, compasses and other position devices.

The model is not an arbitrary construction, but the reflection of the system that is the object of creative activity. The model as presented here will help beginners get a fast start in obtaining useful, creative results. As you learn more and more about TRIZ, you will modify the model and, as TRIZ research continues, more methods and tools will become available for inclusion in the model. The user is encouraged to test and improve the model continuously.

2.8 SUMMARY

The problem-solving model uses five concepts: contradiction, resources, the ideal final result, the patterns of evolution and innovative principles. The diagram shows the relationships among them.

- Contradiction. Solving a problem means removing a contradiction. Contradictions are considered in detail in Chapters 3 and 4.
- Resources. Resources are available, but idle and often invisible substances, energy, properties and other things in or near the system can be used to resolve the contradiction. The mapping of resources is studied in Chapter 5.
- Ideal final result is achieved when the contradiction is resolved. The desired features should be obtained without compromise. The use of the concept of ideality is considered in Chapters 6 and 7.
- Patterns of evolution. Systems evolve according to certain patterns, not accidentally. The patterns can be used many ways to get new ideas and predict the evolution of the system. Five important patterns are presented in detail in Chapter 9.
- Innovative principles give concrete cues for solutions and illustrate what the patterns can mean. The list of 40 innovative principles is studied in the Chapter 10.

Why introduce this model? The model for problem solving connects basic concepts and tools. The integrated system is more effective than the separate parts. It is based on the scientific research of TRIZ and on the feedback from many students of TRIZ over the past two decades.

REFERENCES

1. TechOptimizer™, The Invention Machine Company, http://www.invention-machine.com.
2. Ideation Workbench™, Ideation International Incorporated, http://www.ideationtriz.com.
3. TRIZ Explorer™, Insytec, http://www.insytec.com.
4. The TRIZ Journal, http://www.triz-journal.com. and CreaTRIZ™ **CREAX**, **http://www.creax.com.**
5. Engeström Y., *Learning by Expanding*. Orienta-Konsultit, Helsinki, 1987.
6. Shockley W. The Path to the Conception of the Junction Transistor. IEEE Transactions on Electron Devices, vol. ED-23, no. 7, July 1976, 597

3

CLARIFY THE TRADEOFF
BEHIND A PROBLEM

INTRODUCTION

If a problem exists, clarify the tradeoff behind it. This is the first step in
finding the real problem and good solutions.

We have already said that there are contradictions behind every difficult
problem. The concept of a contradiction is very important. Participants in
problem-solving training often wish for more help with contradiction
analysis, because contradictions are at the core of the most challenging
problems. In this chapter and the next, we will focus on the concept of
contradictions.

Recall the model for problem solving (Figure 3.1). In this chapter, we
analyze the conflict between two features. These are frequently called
tradeoffs because the problem solver "trades" improvement of one feature
against decline in another feature in the hope of finding a solution to the
problem. In Chapter 4, we will study inherent contradictions, when one
thing has two opposite properties. In traditional TRIZ books, tradeoffs are
called "technical contradictions" and inherent contradictions are called
"physical contradictions." We use lay terms because they make sense and
it avoids having special definitions for TRIZ that don't always agree with
everyday language.

1. In this chapter, we discuss in detail why it is so useful to analyze
   tradeoffs. The benefit of TRIZ comes from solving difficult problems
   and that means resolving tradeoffs. To do this, it is necessary to
   formulate these tradeoffs or rewrite the problem in a form that
   makes the tradeoff obvious.

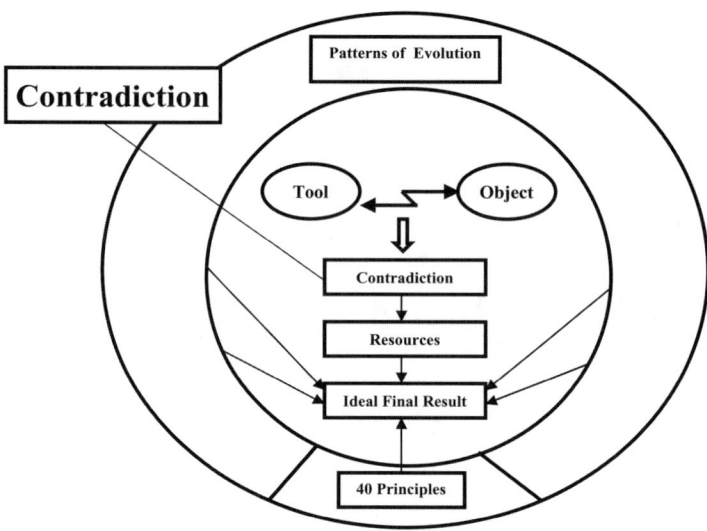

Figure 3.1   Contradiction is the core of a challenging problem.

2. We show how to formulate the tradeoff. The tradeoff appears in the system of the tool and the object. There may be different models of tradeoffs. The system has different features. When we have the problem, we have the tradeoff between features. A simple way to formulate the tradeoff between features is introduced.

3. We consider complex problems that have many tradeoffs. The tradeoff can appear on different system levels and at different times. The selection of the tradeoff is discussed.

4. The last part of the chapter presents five steps for problem clarification.

3.1 WHAT ARE TRADEOFFS AND INHERENT CONTRADICTIONS?

In this and following chapters we will often use terms "tradeoff" and "inherent contradiction." These terms have the same meaning in TRIZ as they do in everyday English.

3.1.1 Tradeoff

When something good happens, something bad happens. Or, when something good gets better, something undesirable gets worse. Some examples of tradeoffs are:

- The product gets stronger (good) but the weight increases (bad).
- Software is made easier to use (good) but versatility decreases (bad).
- The hot coffee is enjoyable to drink (good) but can burn the customer (bad).
- Training gets more thorough (good) but requires more time (bad).
- The faster the automobile airbag deploys, the better it protects the occupant (good), but the more likely it is to injure or kill small people or out-of-position people (bad).

3.1.2   Inherent Contradiction

One thing has two opposite properties. I want it cold but I want it hot. I want it, but I don't want it. There are always inherent contradictions behind tradeoffs — sometimes they are obvious and sometimes they are hidden.

- The product should be thick (to get needed strength) yet should be thin (to be light).
- Software should have very few options for ease of use and should have numerous options to be effective.
- Coffee should be hot for enjoyable drinking and should be cold to prevent burning the customer.
- Training should be lengthy to ensure good learning and should be very short to minimize demands on time.
- The automobile airbag should deploy quickly to save the driver or passenger, yet should deploy slowly to minimize harm to small drivers or passengers.

As the examples show, the terms are easy to understand. If necessary, you can always go back to the examples to recall the definitions.

3.2 WHY ANALYZE TRADEOFFS?

Why are experts in engineering, business and other fields so interested in contradictions today? The most obvious reason is that they need to solve problems. Moreover, the problems they need to solve have an important difference from the ones people dealt with in earlier times.

Mankind has always resolved problems. Homer's *Odyssey* is, actually, a story of problem solving. First, Odysseus had to pass the Sirens and then to sail between Scylla and Charybdis. The Sirens bewitched everybody and there was "no homecoming for the man who draws near them … For with their high clear song the Sirens bewitch him, as they sit there in a meadow piled high with the moldering skeletons of men, whose withered skin still hangs upon their bones." This problem was easy to solve because the goddess Circe gave good instructions: "... to prevent any of your crew from hearing, soften some beeswax and plug their ears with it ... if you wish to listen yourself, make them bind you hand and foot on board and place you upright by the housing of the mast, with the rope's ends lashed to the mast itself."

The problem with Scylla and Charybdis was different. On the route rose two rocks. One was "the home of Scylla, the creature with the dreadful bark ... She has twelve feet, all dangling in the air and six long scrawny necks, each ending in a grisly head with triple rows of fangs, set thick and close and darkly menacing death .... No crew can boast that they ever sailed their ship past Scylla unscathed, for from every ... vessel she snatches and carries off a man with each of her heads."

On the other of the two rocks "... dread Charybdis sucks the dark waters down. Three times a day she spews them up and three times she swallows them down once more in her horrible way. Heaven keep you up from the spot when she does this because not even the Earthshaker could save you from destruction then."

Circe advised: "... you must hug Scylla's rock and with all speed drive your ship through, since it is far better to lose six of your company than your whole crew." Odysseus asked: "Could I not somehow steer clear of the deadly Charybdis, yet ward off Scylla when she attacks my crew?" The goddess gave him a sound berating, called him an "obstinate fool," and continued: "Again you are spoiling for a fight and looking for trouble! Are you not prepared to give in to immortal gods?"

Indeed, Odysseus had no defense against Scylla. The Sirens he passed without problems, following Circe's instructions, but to Scylla he lost six men: "... Scylla snatched out of my ship the six strongest and ablest men. ... I saw their arms and legs dangling high in the air above my head.... In all I have gone through as I explored the pathways of the seas, I have never had to witness a more pitiable sight than that." All citations from The Odyssey are from Rieu's translation of Homer.[1]

For thousands of years, people have accepted that there cannot be better ways to handle difficult problems. Had not the gods themselves warned humans against trying too much? Then, too, new solutions were not needed very often. Many problems were "Siren problems" that could be managed using what they already knew. But this did not apply to all problems. During the same thousands of years, as a result of many cycles of trial and error, qualitative breakthroughs were accomplished: agriculture, clocks, the printing press and other innovations. They clearly didn't fit old thinking schemes. In Homer's terms, once in a while it was possible to sail without any losses past both Scylla and Charybdis.

Today there are more "Scylla and Charybdis" problems than ever before and requirements for the solutions are more stringent. It is no longer acceptable to lose six men, even if the rest of the crew will be saved. Likewise, it is not acceptable to solve a technical problem if the solution causes social problems or to solve a problem for your own customer but cause new problems for other people. And there is no time to wait for the trial-and-error method to eventually come up with a solution.

So, there are two types of problems:

1. "Siren problems," or problems that can be resolved straightfor-wardly using existing rules and instructions.
2. "Scylla and Charybdis" problems or those containing tradeoffs. That is, if you sail too close to Charybdis, you lose the whole ship. If you sail between them, you lose six people. If you sail too close to Scylla, you could lose more than six people. This is the original tough tradeoff.

TRIZ is a new approach that sees contradictions (tradeoffs and inherent contradictions) as sources of development. Resolving the conflicts in a system causes the development of the system. Resolving conflicts is the rationale behind successful inventions and innovations. If you want to move technology forward, you need to understand the conflicts. To do this, one method is to consciously clarify and intensify the conflicts or contradictions. Don't treat them as disorders that should be hidden; treat them as important clues to the solution.

3.3 DEFINING THE TRADEOFF

3.3.1 Tool and Object

The tradeoff arises in a system consisting of a tool and an object. All engineering systems (and many others that would not be considered "engineering" systems) are built from tools and objects. A knife or a saw or a scissors (or anything with a sharp edge) is a tool for working material. A car is a tool for carrying passengers and cargo. A transistor is a tool for switching electrical current on and off. A bee is a tool for collecting pollen.

Knowledge, models and information are also tools. An advertisement is a tool to inform potential buyers. CAD software is a tool for the design of products. An interview is a tool for gathering data. A committee or a team is a tool for making decisions in some companies. The model of TRIZ for problem solving is a tool for improving systems. We can describe the tool and the object using a simple diagram (Figure 3.2).

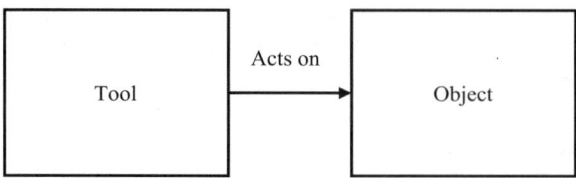

Figure 3.2   Tool, action and object diagram that defines a system.

The Tool is the component that is easiest to change when the problem is resolved. If the problem is how to decrease the wear of a saw that cuts metal, there are many possibilities — change the sharpness or the shape of the teeth, put coatings of tougher metal on the teeth, change the thickness of the blade, add a lubricant during cutting, etc. The metal being cut (the object) most often cannot be changed or can be changed very little.

The Object, on the other hand, gives constraints for the change of the tool. The material being cut limits the choice of cutting tool. Sometimes the object, too, can be changed in some way, for example, many objects can be combined — it may be easier to cut a stack of glass plates than a single thin sheet of glass.

Action means that the tool does something that causes the object to change.

The statement of problem and solving the problem become much easier when the tool and the object that are at the heart of the problem are isolated from the numerous other components of the system. An example: Many gates are locked by a latch mechanism. The mechanism consists of a female part, a male part and a pin. Female and male parts have a hole for the pin. The Figure 3.3 shows how the mechanism works.

The male part is inserted into the female part. Then the pin is inserted into the hole. There is a small clearance between the pin and fixed parts, so that the gate can be easily locked and opened.

Latches of this kind work well on garden gates. But when the same mechanism is used on a big ship, a problem appears. The vessel is constantly moving. Parts of the latch wear and may even fracture. In 1994, the ferry Estonia sank in the Baltic Sea and 852 persons lost their lives. The investigating commission found that the reason for the disaster was failure of a locking mechanism. As the result, a so-called "visor" was lost, water accumulated on the car deck and the vessel capsized. (A "visor" on a ship is a type of gate that resembles the visor in a motorcyclist's helmet.)

If the pin is made easy to lock and open (looser pin), the device gets less reliable. If the reliability is increased (by making the pin tight), opening and locking get more difficult.

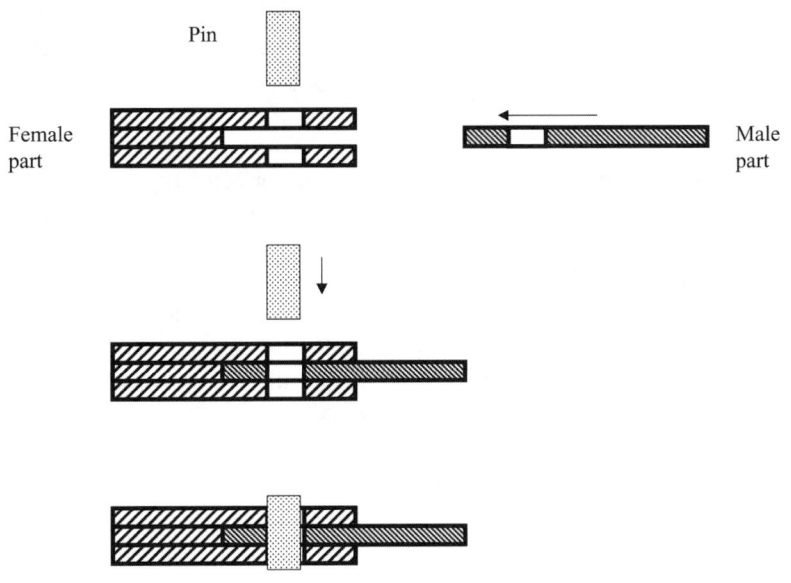

Figure 3.3 Tools and objects in the latching mechanism

To begin studying this problem, we will select the two components that "disturb" each other. Let's imagine a pin or peg and some part with a hole (Figure 3.4).

By limiting our view, the problem has become simpler and better defined. We see only two parts: a pin and a component with a hole. Gate, visor, actuators, electronic control devices and numerous other components have vanished. Using common, everyday words instead of specialized technical terms can help with this step in the analysis.

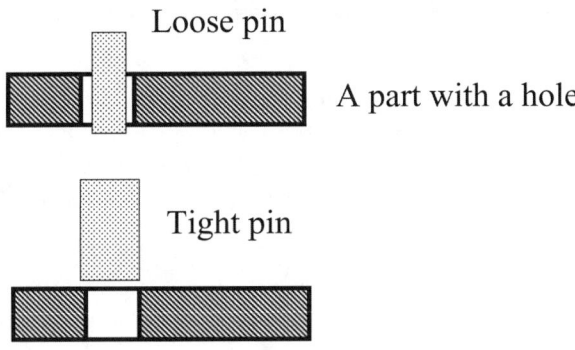

Figure 3.4 Limiting the view in the latching mechanism problem.

A very simple solution to this problem was found and patented in the 1990s (yes, in the last decade of the 20th, not the 19th century). The principle of solution is shown in Figure 3.5. A conical pin replaces the cylindrical pin. The angle of the cone is selected so that the pin is tight enough but will not stick to the latch (The proper angle was 15.4° — details depend on the thickness and the material of the latch and the pin). Excerpts from the patent text: "The locking mechanism according to the present invention assists in eliminating problems arising from the deformation of large gates ... invention includes an elongated guide slot part and an elongated locking part for securing within the guide slot part. The locking part and the guide slot part each include two long sides that are beveled in a similar manner for forming a tight fit.... Moreover, the locking mechanism reduces the amount of force needed to initially open a gate or hatch."[2]

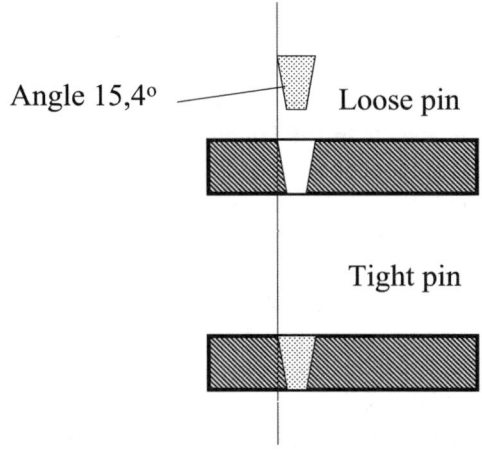

Angle 15,4°   Loose pin

Tight pin

Figure 3.5 The principal solution of the problem in the locking mechanism.

The tradeoff is resolved. A conical pin is easy to insert and remove. In the locked position, it entirely fills the hole and will not wear.

One may ask: What is interesting in this solution? Wouldn't every 10-year-old child find it? If this is your thought, try to answer the following two questions:

1. How was it possible that such a simple solution was not found until the 1990s?
2. What should be done to avoid big losses of time in seeking simple solutions that are badly needed?

In business problems, understanding the purpose of the activity may be even more important than in simple engineering tasks. For example, a company produces boilers, turbines, generators and other energy technology for industry. What is the object of the company's activity? In earlier decades, the answer would be simply to produce and sell boilers and turbines. Today, companies often prefer to say that solving the customers' problems is the real objective of their work. Indeed, the buyer of energy technology usually doesn't need equipment as such, but energy. Do the buyers of energy need just energy? Or maybe they need solutions that help to save energy? This is why companies invest so much time and so many resources to develop mission statements — once the entire company has a common definition of its goal, it becomes much easier to judge any proposed action by whether it advances the goal.

Returning to the discussion of relatively simple systems, it is true (and sometimes confusing) that the same element of the system can be an object or a tool, depending on the problem. If the problem is how to get the ax to cut the wood chunk more effectively, the ax is the tool and the chunk of wood is the object. If the problem is that the edge of the ax is not sharp, the wood is the tool that dulls the blade and the ax blade is the object that is acted on. In some situations, we sell things to the customer and in others, the customer provides data that modifies our actions. The customer can be the tool or the object.

For any particular situation, it takes some work to decide what is the tool and what is the object, but it will make your problem solving easier, so it is worth spending the time. It may be helpful to draw a picture or write the sentence "Tool acts on Object," then substitute the words from your problem.

3.3.2   Tradeoffs Everywhere

It is worth repeating that tradeoffs will inevitably arise in any system. The system consists of the tool and the object interacting with each other. In the top row of Figure 3.6, the straight arrow represents the situation without a tradeoff. Then the tradeoff appears (a flash arrow) and is solved (a big arrow shows a change in the system). The pattern repeats many times as shown in the bottom row.

The tradeoff may appear clearly as a harmful action that accompanies a useful action. In the cutting tool, the edge works the material, but at the same time heats it. The material being worked wears the edge. The car transports people and cargo, but the noise of the car disturbs people and exhaust gases pollute the atmosphere. The transistor produces harmful heat while it does beneficial switching.

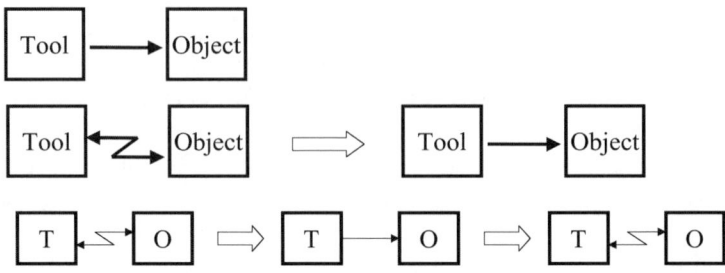

Figure 3.6 A system improves through the resolution of tradeoffs. In the top row, the straight arrow represents the situation without a tradeoff. Then the tradeoff appears (a flash arrow) and is solved (a big arrow shows a change in the system). The pattern repeats many times, as shown in the bottom row.

There are tradeoffs in the system even when there are no visible harmful effects. The electric car is silent and doesn't produce exhaust gases, but the power system occupies a great deal of the space in the car and reduces the load it can carry. No matter what kind of system is used to generate electricity (solar, conventional or nuclear) to charge the batteries, equipment is needed to produce the energy, pollution is generated in the system and additional machinery and energy are used to produce the infrastructure of the power system. Eventually, the batteries must be disposed of, which is a very big source of pollution. Each of these issues gives rise to tradeoffs.

To generalize, the system is bad simply because it exists. To have the capability of acting on the object (useful feature), the tool has dimensions and weight, consumes energy and has other features that engender cost and harm. When the tradeoff is removed, you might think that the perfect system or the ideal final result has been achieved. This achievement, however, is relative and temporary. New tradeoffs appear soon, either in the system itself or in other systems that are affected by it. Every new generation of transistors has occupied less space and consumed less energy than the previous one, but the higher density of transistors on the new chips requires more and more complex systems for connection and for heat dissipation. The electric car eliminates petrochemical pollution, but increases the problem of battery production and disposal.

3.3.3 Different Models of the Tradeoff

To describe the tradeoff between the tool and the object graphically, a variety of symbols are used (see Figure 3.7). The tool can interact in a useful way with one object and in a harmful way with another, i.e., a truck moves cargo but at the same time wears out the road.

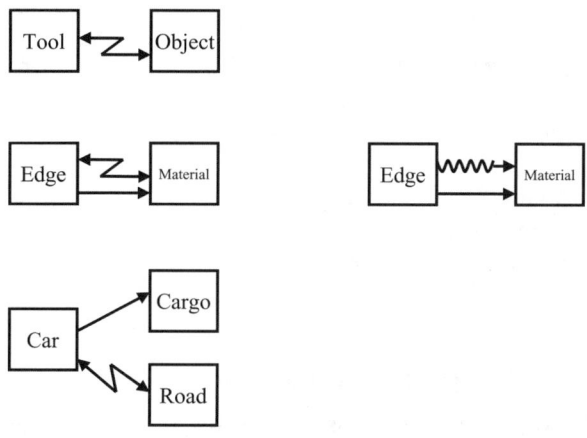

Figure 3.7 **A variety of graphical models are used to show tradeoffs. The flash-like arrow means the tradeoff in general. Sometimes the flash means negative action and the straight arrow a positive one. Edge cuts material, but also heats it. In some TRIZ books, wave-like lines often show a harmful action, particularly in the substance-field (Su-field) method. In some software systems, blue arrows show useful actions and red arrows show harmful actions, so the presence of both colors shows a tradeoff.**

3.3.4 Features

Features and actions can describe tradeoffs. It is simplest to describe features. Most often, the development of a product means the improvement of the features. Usually, the user does not need the product itself, but needs the features. For example:

How can we improve a muffler in the conventional lawnmower powered by an internal combustion engine? First list the features of the muffler:

- Dimensions
- Weight
- Noise absorption capacity
- Ease of manufacture
- Form, outer appearance

How can we thin out carrot seedlings in a small home garden? First list the features of the thinning technology:

- Speed
- Precision
- Ergonomic level

We have spoken of the latch mechanism. How can we improve the pin? List the features of the pin:

- Dimensions
- Form
- Surface quality, manufacturing precision
- Ease of manufacture
- Ease of locking and opening
- Reliability

3.3.5 The Tradeoff Between Features

The tradeoff between tool and object shows where the problem lies. It is useful to express the tradeoff as a conflict between two features. When the velocity of the car increases, safety worsens. A simple diagram, as shown in Figure 3.8, can illustrate the conflict between two useful features.

Often, the improvement of a useful feature is connected with the strengthening of a harmful feature (see Figure 3.9). When the velocity of the car increases, the consumption of fuel also increases.

We can use the features to describe the current state of the system and we can use them to describe the system that we want. If there is a conflict between the features or a tradeoff, we can use TRIZ to remove the tradeoff and create the new, improved system.

Sometimes, the analysis of functions and actions can give valuable additional information about tradeoffs. This is the case if functions can be expressed clearly enough. Often, however, one cannot find verbs and nouns that can satisfactorily describe functions of even simple everyday systems. What is the function of a water tap? Is it to control water flow? "Control" is an extremely general verb that does not tell how to detect the function of delivery of water. "To stop water" or "to regulate water" both define only one half of a function. In contrast, one can easily list essential features of a water tap: water flow, reliability, ease of use, decorative appearance and others. Therefore, we will use features rather than functions for most of our examples.

When people describe problems, often only the drawback is mentioned instead of the tradeoff. Insufficient safety or excess consumption of fuel

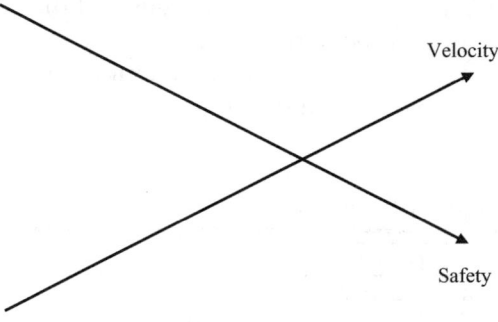

Figure 3.8   A conflict between two useful features.

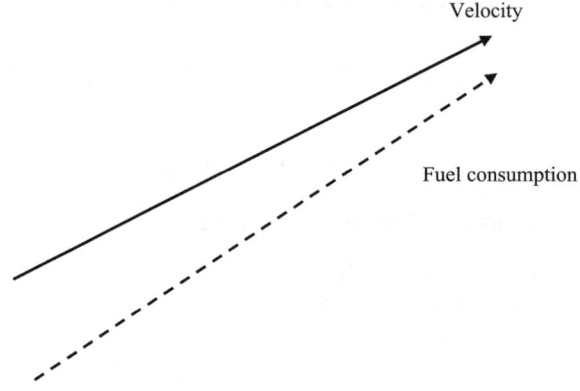

Figure 3.9   A conflict between useful and harmful features Here, the dotted line in the figure means a harmful feature.

is identified as the problem with a car and the tradeoff (fuel consumption increases when speed increases or safety decreases when speed increases) is not mentioned.

Sometimes, desired features are formulated instead of the tradeoff. Safety and fuel economy are described as "tradeoffs" without specifying what feature is diminishing when safety or fuel economy improves. The problem solver tries to jump directly to the solution.

Analysis of both sides of the tradeoff is essential. It is not academic hair-splitting. When the connection between good and bad features is expressed, some features of the solution begin to appear. We know what tradeoff the solution should remove, although we don't yet know how.

In real situations, problems contain many possible pairs of tools and objects and many tradeoffs. We recommend trying the exercise in Table 3.1

to see the multiplicity of tradeoffs. It is suggested that you fill in the table with examples from your personal life and from your business experience. Which components and which tradeoffs should you select to get a good solution to your problem? In the following section, we will consider the selection of tradeoffs.

Table 3.1   Examples of Tradeoffs

When THIS gets better	THIS gets worse
The size of the warehouse increases	The accuracy of the inventory decreases
My family is happy with a vacation	It takes too much time from my work

3.4 AN ABUNDANCE OF TRADEOFFS

3.4.1 Where Does the Tradeoff Appear?

The problem statement is like the answer to a journalist's questions: who, what, where, when, why and how. Answering the questions may reveal the tradeoffs. The level of the system is important, too. The system contains some parts of lower level and is itself a part of the higher-level system or macrosystem.

The muffler of the lawnmower is a system containing at least two subsystems: a casing and porous absorbing material. The muffler itself is a part of the lawnmower. The lawnmower, further, is a part of a park system or a garden system that contains grass, dirt, the gardener and other systems.

There are tradeoffs and solutions on all levels:

■ The level of the muffler. The lower the level of noise, the thicker the layer of porous material. One possible way to decrease noise without thickening the muffler is using noise cancellation. An active noise-control device generates sound waves whose peaks correspond to the valleys of the undesired sound and vice versa. A system like this is used in some cars.

■ The level of the lawnmower. This system contains casing, engine, exhaust tube, muffler and other parts of the lawnmower. If noise is decreased using the muffler, the lawnmower becomes more complex.

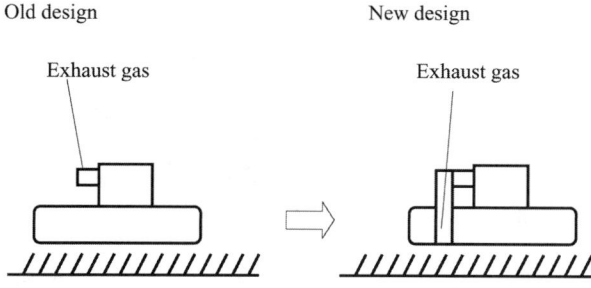

Figure 3.10   Grass working as sound-absorbing material.

One obvious problem statement is how to change the casing so that it also does the job of the muffler.

■ The level of a garden (or yard or park). The system contains at least the lawnmower and grass. The number of problems and solutions increases. The grass can be used two different ways:

A. A Finnish inventor used grass. He turned the exhaust tube down into the grass (see Figure 3.10). Grass worked as sound absorbing material. Noise decreased considerably. Hot gas also dried the grass so that it didn't stick to the lawnmower.

B. When the appearance of the garden is improved using the lawnmower, the system gets more complex. If the grass does not need to be cut, we won't have to worry about the noise from the lawnmower. Grass that grows to a certain height and then stops has been developed and is being tested for commercial use. The Japanese have hundreds of years of experience developing moss gardening. Moss carpet doesn't need any cutting. In trailer parks, people frequently paint rocks green and have no grass. Sports centers often use artificial grass such as Astroturf.[8] Clearly, the best solution to the problem requires knowing what the use of the system will be.

It is useful to formulate problems and tradeoffs on more than one level (see Figures 3.11 and 3.12). The selection of the tradeoff depends on the constraints and conditions of the problem. If you need a quick and cheap solution, it may be best to begin with mechanical changes to the lawnmower or the muffler. If there are more resources available, solving problems on the micro- and macrolevel can be more exciting and can suggest changes that will influence the whole industry.

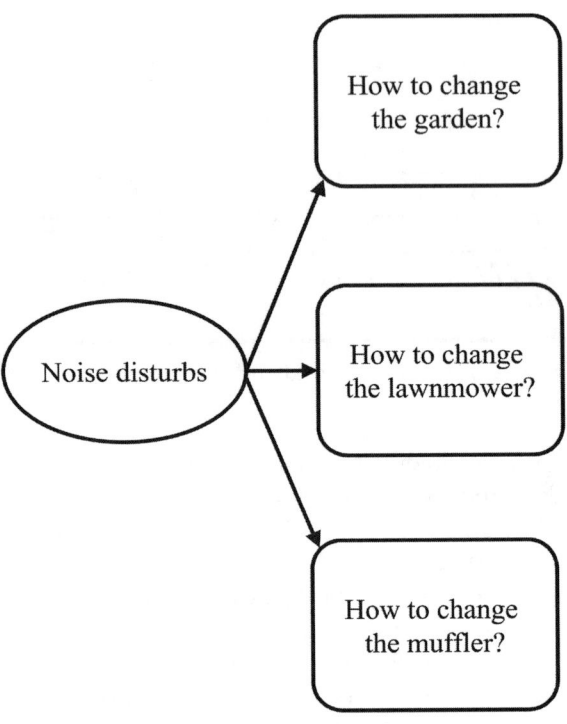

Figure 3.11   Formulate problems on more than one level.

There is no way to guarantee a perfect problem statement. But, if you start with the simplest, most obvious statement, you will be able to improve it later. It is important to start.

3.4.2   When Does the Tradeoff Appear?

Problems can appear at different times in the life of the system. Another gardening example can illustrate this. Carrots are cultivated in a small home garden. The initial problem is how to make the thinning of carrot plants easier. Thinning is necessary so that there will be enough space between the plants as they mature to allow each plant to get enough nutrients and water. Before hurrying to solve this problem, it is useful to consider the stages of cultivation. For simplicity, we'll look at only two stages:

■ Seeding
■ Thinning

Both stages contain their own problems and tradeoffs.

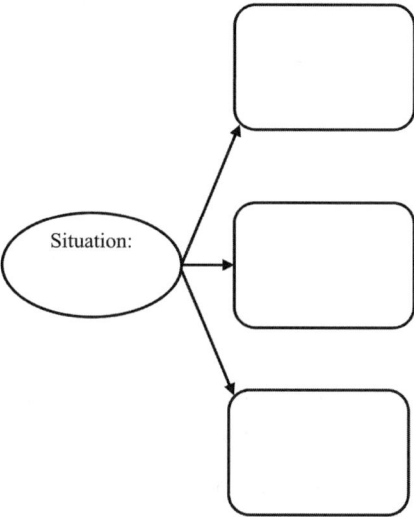

Figure 3.12 Exercise. Formulate problems on more than one level.

- Seeding. If every seed is planted precisely in the right place, no thinning is needed. If one plants seeds very simply by hand, much time is needed. Time can be saved using seeding equipment, but then the drawback is the existence of the equipment, its cost, storage, maintenance, etc.
- Thinning. The same tradeoff as in seeding: thinning can be mechanized some way and time saved, but the system becomes complex.

There is one beautiful solution that makes seeding simple and thinning totally unnecessary. Seeds are fixed on a biodegradable tape. The gardener places the tape in the furrow. The plants grow in exactly the right places (see Figure 3.13). You can, yourself, make a seed tape from paper towels, white glue and small seeds. See, for example, URL http://lancaster.unl.edu/hort/youth/seedtape.htm.

Sometimes a good solution is found in another stage of the process. That's why it is useful to formulate tradeoffs in different process stages. In this example, the problem of the thinning stage can be eliminated by a change in the seeding stage.

3.4.3 Nine Screens

The location of the problem in time and on the system level can be illustrated by a simple table containing nine screens (sometimes called

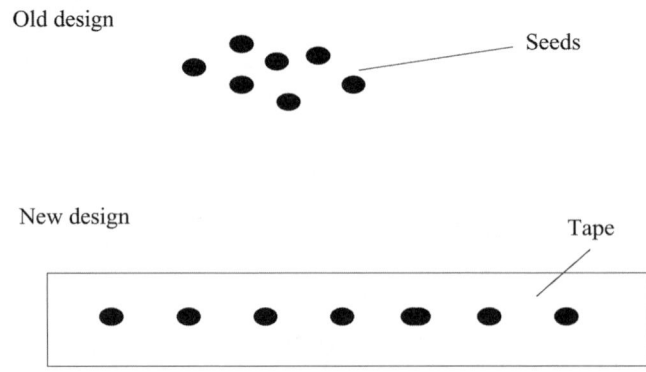

Figure 3.13 Seeds fixed on a biodegradable tape.

windows, boxes, the "tic-tac-toe" method or the System Operator.) We show the examples of the lawnmower and the cultivation of carrots on the nine screens (see Figure 3.14)

The table helps to see the problems more clearly and sometimes suggests solutions very directly. In the carrot example, we considered present and past. How about the future? If the carrot row is not thinned, roots will be small. If we can find a use for "minicarrots," thinning is not needed. How about macrolevel? A tape with seeds can be considered a macrolevel system. How can this system be improved? For example, the tape could contain fertilizer.

Altshuller called the table "screens of talented thinking." Usually, people see one screen: a system at present. A talented thinker sees at least nine screens: system, a macrosystem, a microsystem and all three levels in past, present and future. Figure 3.15 presents a template for the use of nine screens.

This is the simplest method for using the nine screens. The system of screens is an independent set of tools that can be used many different ways — to enhance your understanding of a problem and to help you expand the areas in which you can look for solutions. You can obtain many more ideas about using the nine screens from four articles by Darrell Mann in *The TRIZ Journal*.[3]

3.4.4 How to Decide Whether to Develop the System or Remove It

Sometimes, it may be simpler to remove the system than to develop it. In his historic work on brainstorming and idea generation, Alex Osborn suggested asking the questions: "What can we eliminate?... Suppose we leave this out....Why not fewer parts?"[4] Usually, however, it is not possible simply to leave out a part or a stage of a process. The elimination of a part

	Past	Present	Future
Macro-level		Lawn-mower	
System		Muffler	
Micro-level		Porous substance	

	Past	Present	Future
Macro-level			
System	Carrot: seeding	Carrot: thinning	Carrot: consumption
Micro-level			

Figure 3.14   Examples of modeling by nine screens.

or operation means, in most cases, that some useful features and functions will disappear. By suggesting the removal of a part or a system or a process, we have created a tradeoff. If we can formulate the tradeoff explicitly, then resolve the tradeoff so that the system will have fewer parts and operations and more useful features, we will have significantly improved the system. Conventional thinking often stops before considering this kind of tradeoff.

What is the new perspective that TRIZ gives here? Instead of the suggestion: "Suppose we leave this out," we use new questions: "What are the good and bad features of this component or operation? What gets better and what gets worse, if we leave this out?" Examples:

	Past	Present	Future
Macro-level			
System			
Micro-level			

	Past	Present	Future
Macro-level			
System			
Micro-level			

Figure 3.15 Exercise. Give examples from your personal life or your business situation using the nine screens.

- What is good and what is bad about the spare tire in the car? It increases reliability, but at the same time occupies space. In some cars, the normal tire is also the "spare." It is designed so that, after a puncture, the car can be driven cautiously to the nearest repair shop.
- In the beginning of this book, we gave the example of the pipe with a T-joint. T-fittings are needed to make a complete pipe system (+), but they increase the number of parts (–). A collared pipe is the solution that provides the function of the T-joint without the complexity.

- Operations can also be removed. The harvester cuts and reaps at the same time, although, years ago, cutting and reaping were separate operations. Digital printing removes typesetting. On-the-job training removes classroom training.
- Storage operations and warehouses, once considered necessary, have been practically eliminated in many industries, such as electronics. Products are manufactured strictly on demand, then delivered directly to the point of use. But there are also industries where storage problems persist. Much food is wasted in markets. Distributors keep extra food to ensure that buyers can find what they need. But, some of the food will be left in the warehouse. How would you solve this problem?

Removing the component or operation is not always possible. Often, some other solution for the tradeoff is needed, but this method, thinking about what would happen if we removed the component or operation, helps to find the real problem.

3.4.5   How to Identify the Right Problems to Solve

An old proverb reads: "It is more important to do the right things, than to only do things right." Unfortunately, repeating this and similar statements helps very little. Problem selection in the past has often been the expert's subjective choice.

TRIZ is revolutionizing the technology of problem solving and it is also changing the methodology of developing the problem statement. Instead of subjective or arbitrary formulations, using TRIZ we have a precise definition. A real problem contains tradeoffs.

The analysis of tradeoffs is only one of the tools in the toolkit that TRIZ provides for developing a precise problem statement. All the tools presented in this book can be used for both problem finding and problem solving. You can apply the tools one at a time as you read each chapter, but you will enhance your problem analysis and problem-solving skills by using all the tools together.

In this section, we focus on tradeoffs. Finding the best problem to work on means selecting a relevant conflict from many tradeoffs.

First, consider the available time and resources. In principle, there are many alternatives. In practice, the choice is much more limited. In the lawnmower example, the most dramatic improvement is to get nice looking grass without a lawnmower. This might be the best solution, but it requires a large investment of research money over a considerable period of time, then another large investment to introduce the new groundcover to the market. If the lawnmower company needs a quick or a cheap improvement, it might be better to examine the muffler and the elements of the system that interact or could interact with the muffler. The same can be said of making choices for improvements at different stages of a process. Sometimes one can solve the problem by going to the past (example of carrot cultivation) or to the future and sometimes not.

Second, the problem appears between certain components. That simplifies the choice of the problem. You simply select the components touched by problems. The pin and the part with a hole "disturb" each other in the latch problem. That's why they are selected. A gate and actuators, for example, are omitted. They are not connected with harmful interactions or features.

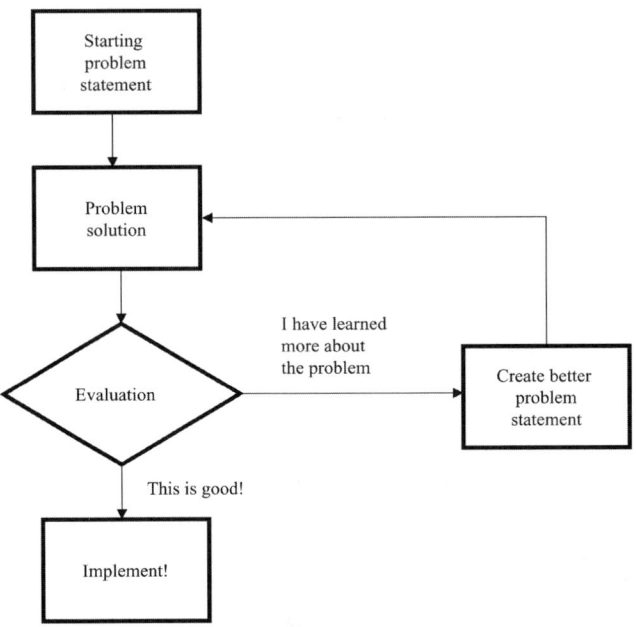

Figure 3.16 Flowchart for getting started in problem solving. In a flowchart, a box is an action and a diamond is a decision. The arrows show the order in which things are done. This flowchart describes the method of starting with a preliminary problem statement, then developing better problem statements as you learn more about the problem.

These recommendations, however, don't guarantee the right choice for the problem. If I can find a groundcover that doesn't need cutting, the right task is to remove the lawnmower from the system. If I cannot, the right task is to improve the lawnmower. That is, to formulate the right problem, we should know the solution. But, to get the solution, it is very important to state the right problem. Behind popular phrases like, "A problem well stated is half-solved," lies a silent assumption that one can state the problem well. This assumption is wrong. No expert in the world can state the problem exactly at first. Is this a deadlock? What, then, should the TRIZ practitioner do?

There is a well-proven, practical approach. State the problem that seems to be most reasonable on the basis of existing knowledge and try to solve it. During the attempt to get the solution, new obstacles as well as new opportunities will appear. This new knowledge allows us to modify the problem statement. This kind of generation of new ideas may remind us of the exploration of unknown continents in the past or of space research today. We don't know what's out there, but we have some basis for the assumption that it is worth going. Later explorers (or the first explorers on later trips) will have better ideas about what they should look for, so they can bring better tools or people with more specialized skills, etc. See Figure 3.16.

3.5 FROM THE PROBLEM TO THE TRADEOFF

Let's summarize the points in this chapter. There are five steps for the clarification of a problem:

1. Describe pairs of tools and objects and the action that links them. Select one pair. Explain why you picked this tool and object.
2. Describe features and conflicts between them. Select one tradeoff.
3. Explain why you identified this tradeoff.
4. Describe the tradeoff graphically and in words.

The following tables and figures contain the summaries of some examples: noise problem in a lawnmower (Table 3.2 and Figure 3.17), cultivation of carrots (Table 3.3 and Figure 3.18) and problems in the latching mechanism (Table 3.4 and 3.19). Table 3.5 is a template for the study of own problems.

3.6 SUMMARY

- Why study contradictions (tradeoffs and inherent contradictions)? The two kinds of problems are Siren problems and Scylla problems. You don't know which kind of solution to use if you don't know which kind of problem you have. Scylla problems contain contradictions. Resolving contradictions in a system causes the development of the system.

Table 3.2. Constructing the Model of Tradeoffs Using the Lawnmower Example

1. Describe pairs of tools and objects.
   - Muffler and noise (or air that vibrates)
   - Noise and a person (disturbed by noise)
   - Engine and exhaust gas (engine moves gas)
   - Etc.

2. Select one pair. Explain why just this tool and object are selected.

We select the pair "muffler and noise" because we want to select a limited problem. We want remove the harmful feature with minimal changes in the system.

3. Describe features of the selected system of a tool and an object. Describe conflicts in this system.

Features: dimensions, weight, absorption capacity, ease of manufacture, form, etc.
Conflicts between features:
   - When noise absorption improves (+), the dimensions of a muffler increases (−)
   - When noise is suppressed (+) the number of parts increases (muffler needed) (−)
   - ■

4. Select one pair of conflicting features. Explain why just this tradeoff is selected.

We select the formulation: When noise is suppressed (+), the number of parts increases (muffler needed) (−).
The reason: A limited problem, very big changes not needed, at the same time the goal is more than simple optimization of the size.

5. Describe the tradeoff graphically and in words.

Graphically: see Figure 3.17.
In words: When noise absorption capacity improves, dimensions (of the muffler) and number of parts (the muffler is needed) are increased, too. If the system is simplified (or muffler made smaller or removed totally), the noise absorption capacity is lessened or lost.

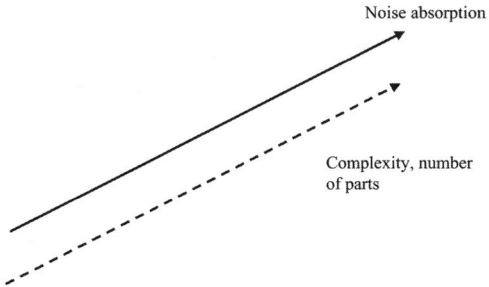

Figure 3.17   Tradeoff in the noise suppression problem.

Table 3.3   Constructing the Model of Tradeoffs: How To Cultivate Carrots

1. Describe pairs of tools and objects
    - Hand and seed
    - Hand and carrot
    - Seeder and seed
    - Etc.

2. Select one pair. Explain why just this tool and object are selected.
    - Hand and seed. The constraints of the problem allow us to make changes easily in the early stages of the process. The simplest system is selected.

3. Describe features of the selected system of a tool and an object. Describe conflicts in this system.
Features: accuracy, speed, convenience
Conflicts between features:
    - The more accurately the carrot is seeded (by hand), the lower the speed
    - The higher the speed when seeding precisely (by machine), the more complex equipment

4. Select one pair of conflicting features. Explain why just this tradeoff is selected.
We select the first formulation: "The more accurately carrot is seeded (by hand), the lower the speed." The reason: Try the simplest case first.

5. Describe the tradeoff graphically and in words:
Graphically: see Figure 3.18.
The more accurately the carrot is seeded (by hand), the lower the speed.

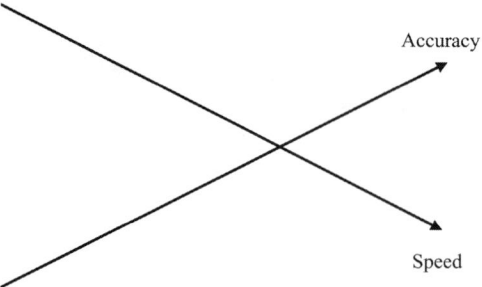

Figure 3.18   Tradeoff in the carrot cultivation problem.

Table 3.4   Constructing the Model of Tradeoffs: How to Improve the Latching Mechanism

1. Describe pairs of tools and objects.
- Female and male parts
- Pin and a part with a hole
- Etc.

2. Select one pair. Explain why just this tool and object are selected.
Pin and a part with a hole. The problem, wearing, occurs between a pin and a part

3. Describe features of the selected system of a tool and an object. Describe conflicts in this system.
Features: Dimensions, form, surface properties, manufacturing precision, ease of use, reliability.
   Conflicts:
- If the pin is made easy to lock and open (loose pin), the device gets less reliable. If the pin is made tight, opening and locking get difficult.
- If the pin is machined very precisely so that is both more reliable and easy to use, manufacturing gets complex.

4. Select one pair of conflicting features. Explain why just this tradeoff is selected.
We select the first formulation:
- If the pin is made easy to lock and open (loose pin), the device gets less reliable. If the pin is made tight, opening and locking get difficult.
The reason: The system that doesn't need precise machining is simple.

5. Describe the tradeoff graphically and in words:
Graphically: see Figure 3.19.
If the pin is made easy to lock and open (loose pin), the device gets less reliable or wearing gets worse.

Table 3.5   Construct the Model of Tradeoffs: Consider Your Own Problem Situation

1. Describe pairs of tools and objects.

2. Select one pair. Explain why just this tool and object are selected.

3. Describe features of the selected system of a tool and an object. Describe conflicts in this system.

4. Select one pair of conflicting features. Explain why just this tradeoff is selected.

5. Describe the tradeoff graphically and in words:

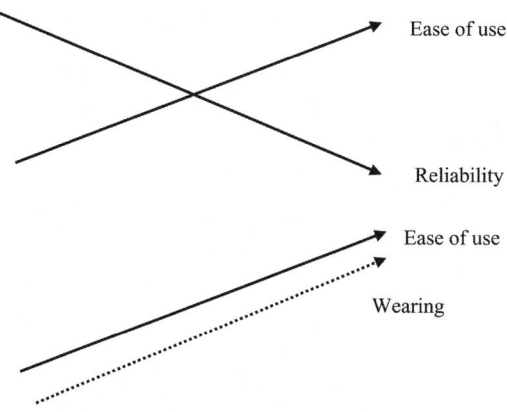

Figure 3.19   Tradeoffs in the problem of the latching mechanism. Two pictures illustrate that the same conflict can be visualized different ways.

- To find the tradeoff, first select the tool and the object. Then, describe the important features of the selected system of the tool and the object. Select the pair of conflicting features that best characterize the problem. Use different verbal and visual models for describing contradiction.
- There are many tradeoffs. Ask where and when the contradiction appears. The scheme of nine screens or windows helps to answer the questions when and where. Decide whether to develop the system or remove it by formulating the contradictions that appear if a component or operation is removed. How can you find the right problems to solve? Try a variety of models and formulations of tradeoffs first, to state the problem that is most reasonable and then modify the problem statement, if necessary.
- Go from the vague problem situation to the clearly formulated tradeoff: describe pairs of tools and objects and the action that links them. Select one pair. Describe features and conflicts between them. Select one tradeoff. Describe the tradeoff graphically and in words. Sometimes the formulation of the tradeoff gives you the idea for the solution. Sometimes you need other tools, which will be presented in later chapters of the book. Even if you have a solution that you like as a result of this method of clarifying the tradeoff, it would be a good idea to use the tools presented in later chapters in the book to develop more solutions.

REFERENCES

1. Homer, *The Odyssey*. Rieu´s Translation, Penguin Books, London, 1991, 180.
2. Lahtinen, M.U.P. and Holtta, P.J., U.S. Patent 5,875,658, 1999.
3. Mann, D. System Operator Tutorial 1-4, *The TRIZ Journal*, September 2001, November 2001, December 2001, January, 2002, www.triz-journal.com.
4. Osborn, A.F., *Applied Imagination*, New York, Charles Scribner's Sons, 1963, 267.

4

MOVING FROM TRADEOFF TO INHERENT CONTRADICTION

INTRODUCTION

The crazier the conflict you imagine, the better solution you get. In previous chapters, we briefly defined two different types of contradictions: tradeoffs and inherent contradictions. We have considered tradeoffs in detail. In this chapter, we learn how to to move from the tradeoff to the inherent contradiction, and how to intensify the contradiction.

Why, in addition to tradeoffs, do we need another contradiction? First, the formulation of the inherent contradiction helps to get the best solutions because it helps find the key problem. There are always many problems and tradeoffs, but not all problems are equally important. One problem is key. The solution of this problem leads to the solution of others. Recall the fast-food example in the beginning. Old drive-in restaurants had many problems and tradeoffs: slow service, high costs, uneven quality, not always the best reputation, and others. The whole bundle of drawbacks resulted from one inherent contradiction: the restaurant needed many service people, and it needed very few service people.

The second reason is that the formulation of the inherent contradiction includes one element of the good solution. The inherent contradiction should be removed. The self-service concept clearly removed the contradiction "many–few service people." There are no service people. At the same time, we can say that there is one service person to each customer — himself or herself.

Third, if we can focus on one key problem, we can present problems and their solutions much better to all possible customers, as well as management, colleagues and partners. The formulation of the inherent contradiction is rewarding in many ways.

In this chapter, we first consider how to formulate the inherent contradiction. Instead of the two conflicting features that we worked with when we used the tradeoffs, we will have only one feature with incompatible values (big–small, many–few, long–short, much–little). Second, we describe how to intensify the contradiction further. Often the contradiction can be made sharper, even more bizarre, which will help point the way to good solutions. The third section presents examples of how to go from the visible drawback to the intensified inherent contradiction.

4.1 HOW TO FORMULATE THE INHERENT CONTRADICTION

So far, we have described contradictions between a tool and an object or, generally, in the system of the tool and the object. But real problems frequently are not simple, and there are conflicts within each object, as well as between them (see Figure 4.1).

To act on the object, the tool should have certain weight, dimensions, energy consumption and other features. At the same time, we want to minimize weight, size and energy losses. The tool should have different, incompatible values of a single parameter. Examples of typical conflicts:

- Big–small. To be comfortable, the house should be big and, to decrease the comsumption of material and energy, the same house should be small.
- Heavy–light. Strength, reliability and safety require that a vehicle should be heavy, but energy economy requires that the same vehicle should be light.
- High energy consumption–low energy consumption. Machinery and tools need to use energy to work on objects, and they should use as littele energy as prossible.

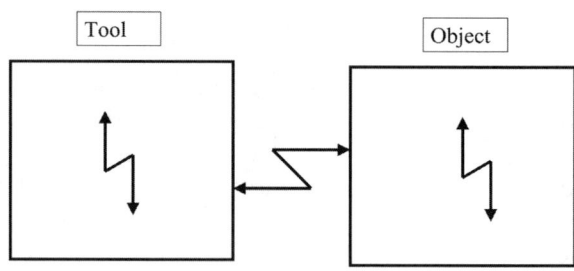

Figure 4.1 Conflicts within and between components. Tradeoffs (between components) may be difficult to solve directly because the components should have two opposite properties.

Generally, to improve feature A, an element should have a certain property, but to improve feature B, the same element should have the opposite property.

Just the formulation of the inherent contradiction, incompatible values of the same parameter, can guide us to ideas for solutions, as we can see in our examples. In the lawnmower noise problem we had a tradeoff: if noise absorption improves, the size of the muffler increases. We can easily formulate the inherent contradiction "big muffler–small muffler."

Big Muffler–Small Muffler

In the example of carrot cultivation, we have a tradeoff: The more precisely we seed carrots (by hand), the more time is needed (the slower the speed). Here, a little bit more thinking may be required to find the inherent contradiction. If we plant a lot of seeds, we have a tradeoff between precision and speed. If we plant only two or three seeds, this tradeoff disappears, but we will have another one: we can seed precisely and not much time is needed, but the crop will be small. So we have the inherent contradiction.

Many Seeds–Few Seeds

In the seeding problem, the conflict is connected with the object, not with the tool. Objects often contain contradictions such as many objects–few objects. There can also be an inherent conflict with the tool (the person planting the seeds.) The planter wants high speed for a large crop, but low speed for precision planting, so the work of thinning the carrots later can be avoided.

In the example of the pin in the latching mechanism, we had a tradeoff. If the pin is made easy to lock and open, wear gets worse. The inherent contradiction is loose pin–tight pin, or big clearance–small clearance (between pin and part).

Big Clearance–Small Clearance

Conflicts don't have to deal with physical objects. In the example of customer service, we need training so that employees can offer good service. But, while employees are attending the training class, they are not doing the work that their company and their customers need done. An inherent contradiction behind this tradeoff is long training–short training. At the extreme, this becomes continuous training–no training or excellent service (after the class)–no service (during the class.).

Long Training–Short Training

Using water to fight fires is a technical example that is familiar to everyone. But, the more water used, the more equipment needed. And, if the fire is put out, large amounts of water can damage the parts of the building that were not damaged by the fire. In some cases, the water does more damage than the fire. The inherent contradiction is much water–little water.

Much Water–Little Water

Not long ago, a simple and interesting solution was developed. Water is atomized and sprayed in very small microdrops, or mist. Much less water is needed to extinguish the fire and there is much less damage caused by the water itself. Of course, there is the new contradiction in the complexity of the atomizing equipment. That contradiction is partially eliminated by virtue of decreased water flow and high pressure — the size of tubes and other equipment can be decreased. The same volume of water gives 8000 times more drops than a conventional sprinkler and 200 times more than low-pressure mist equipment (Figure 4.2). Because the extinguishing efficiency depends mainly on the number of drops, pipe sizes can be radically decreased. For more information, see Web page http://www.hi-fog.com.

4.1.1   Present and Absent

One important group of inherent contradictions is made up of situations in which the object should be present to get some useful feature, and absent to keep the system simple. In Section 3.4.4, we presented examples of tradeoffs that arise if we want to remove parts or eliminate operations from a system. Now let's reexamine these tradeoffs and look at the inherent contradiction behind each one:

- Tradeoff: A spare tire in the car increases reliability (+) but occupies space (–).
  - Inherent contradiction: A spare tire should be present *and* absent. One beautiful solution is to use the normal tire as the "spare" by making it able to survive a blowout (inherent contradiction solved), rather than making the separate spare tire smaller, as in some cars (compromise).
- Tradeoff: If the complexity of a plumbing system is improved and liquid is distributed to more points (+), a greater number of T-joints is needed (–).
  - Inherent contradiction: A T-fitting should be present *and* the same fitting should be absent

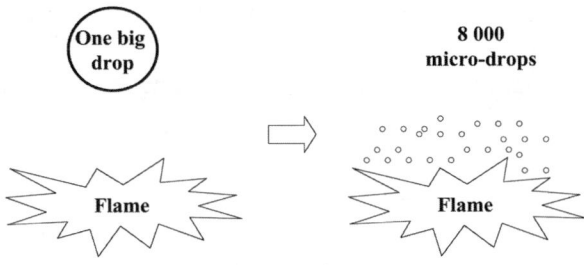

Figure 4.2   The conflict much water–little water resolved.

- Tradeoff: If a plentiful supply and good assortment of food are always available in the warehouse (+), considerable food will be wasted (–).
  - Inherent contradiction: Extra food should be present *and* extra food should be absent.

Do you have some ideas about how to resolve these contradictions? Write them down in Table 4.1 before you read the rest of this paragraph. Don't look at the following suggestions until you have written your own. There are many ways to solve each of these problems, and which idea is best will depend on the circumstances.

1. How can we eliminate the spare tire? Some truck companies don't have any spares in the vehicle. When a blowout happens, the driver calls the service center and gets a new tire. It is cheaper to wait a little for the new tire than to carry an expensive extra tire all the time.

2. How can extra food can be present and absent? Maybe you can find buyers who will buy extra food near the last day it can be sold at discount price. The seller could recoup some of the cost of the food and prevent waste. Would some "food-on-demand" concept work? In that scenario, all food could be sold at normal price. Maybe extra food could be presented to a children's hospital or to another charity. In this case, it stops being "extra" food and becomes socially useful, like the Second Harvest program in many U.S. cities.

Tradeoffs and inherent contradictions appear in all areas of endeavor — engineering, commercial, organizational, educational and social situations, etc.

In the 1950s, Altshuller wrote that finding and resolving contradictions is essential in problem solving. During the past few decades, the same contradictions have been discovered in science, engineering, and business situations that, on the surface, appear to be very different from each other.

Table 4.1   List Your Ideas about How to Resolve Contradictions

Problem	Ideas
Tire reliability	
Plumbing parts	
Extra food	

Table 4.2   Adding Examples of Inherent Contradiction

Study examples of the inherent contradiction. Add three more examples of your own. Try one each from your business life, your personal life, and your community.

Peters and Waterman wrote the bestseller *In Search of Excellence* in 1982. At least one chapter in the book is absolutely true today — perhaps more true today than it was 20 years ago. In Chapter 4, Managing Ambiguity and Paradox, they write: "Most important, we think the excellent companies, if they know any one thing, know how to manage paradox."[1]

Let's go from business to dramaturgy. Syd Field writes in *The Screenwriter's Workbook*: "Drama is conflict; without conflict there is no action, without action, no character, without character, no story, without story, no screenplay".[2]

It is useful to collect your own examples of inherent conflicts (see Table 4.2).

4.2 HOW TO INTENSIFY THE INHERENT CONTRADICTION

We can sometimes make the internal conflict we just formulated much stronger. This will make the problem seem unsolvable — you might think this is a bizarre technique, but it can lead to great solutions. Think of solutions such as:

■ The aeroplane is heavier than air (more dense) yet, in some sense, lighter (thanks to the lifting force).

- The transistor can be a conductor and an insulator at different times.
- Biodegradable materials are present and exist (at certain times) and become absent and nonexistent (at other times).

Let's return to examples studied earlier. The contradiction can be intensified:

- Instead of the small muffler we require, try the idea of no muffler at all.
- Instead of the few seeds we require, try the idea of a single seed.
- Instead of the small clearance we require, try zero clearance.
- Instead of the short training we require, try no training time at all.
- Instead of decreasing the amount of water to fight the fire, try no water at all.

Intensifying the contradiction is the key to the solution. When the grass is used to absorb noise, we have either no muffler or a very big muffler. When many seeds are fixed on tape, we have one seed planted at a time. A conical form pin has no clearance during work, and sufficient clearance when one needs to close or open the mechanism. Training embedded in work may be long lasting, although classroom training may not even take place. Water mist has many properties of a gas. It is practically dry, that is, it doesn't cause water damage, yet it is wet enough to suppress fire. There is a qualitative difference between "little water" and "no water."

These examples show how the TRIZ technique of forcing the contradiction to the extreme can be a technique for breakthrough. Analyzing the extreme contradiction helps one break out of conventional thinking into the realm of ideas that get rid of the contradiction, and solve the problem.

If the wildness of extreme contradictions makes you uncomfortable, remember that they are only needed when new ideas and solutions are needed. They are supposed to shake you up, to get you to examine possibilities outside the system that caused the problem in the first place. If you have an existing solution to the problem that needs minor improvement, go ahead and improve it — don't go to the extreme. To get comfortable with extreme conflicts, try the exercise for intensifying conflict in Table 4.3.

For further study of inherent contradictions, you can find material in other books. Altshuller's *And Suddenly the Inventor Appeared*[3] contains a short introduction to the concept of the physical contradiction. The same topic is considered in Savransky's book[4] beneath the term "physical point contradiction."

Table 4.3   Examples of Intensified Conflict

Study examples of intensified conflict. Add three more examples of your own. Try one each from your business life, your personal life, and your community. You can intensify, if possible, conflicts you have formulated in Table 4.1, or add totally new examples.

4.3 EXAMPLES

Let's review how we applied the idea of the hidden inherent contradiction and the intensified contradiction to our examples (Tables 4.4–4.6).

Table 4.4   Reducing Lawnmower Noise: Intensifying Contradiction

Modeling steps	Example: Reduce lawnmower noise.
Visible drawback	The lawnmower is too noisy
Tradeoff: the conflict between two features	When noise absorption improves, the size of muffler and number of parts increase
Inherent contradiction	Big muffler–small muffler
Intensified inherent contradiction (if it can be intensified)	Big muffler–no muffler

Continue the study of your own problems. Summarize the intensification of the contradictions. Does the intensification give you any new ideas about solutions? See Table 4.7.

Altshuller published the first article on TRIZ in 1956 with his friend and colleague Shapiro. (The paper wasn't published in English until 2000.) Defining the critical contradiction and determining the immediate cause or contradiction were named as essential stages of problem solving.[5] Later, in the 1970s, Altshuller expanded the definition from one type of contradiction to two: tradeoffs and inherent contradictions (technical and physical contradictions by the old terminology). If you are learning TRIZ now, you can start with these important concepts and save 20 years.

Table 4.5   Cultivating Carrots: Intensifying Contradiction

Modeling steps	Example: Make it easy to grow carrots
Visible drawback	Thinning of carrots is an arduous job
Tradeoff: the conflict between two features	The more precisely carrot seeds are planted, the slower the speed
Inherent contradiction	Many seeds–few seeds
Intensified inherent contradiction (if it can be intensified)	Very many seeds–one seed

Table 4.6   Improving Latching Mechanism: Intensifying Contradiction

Modeling steps	Example: Improve pin-type latch
Visible drawback	Latching mechanism wears and can even fracture
Tradeoff: the conflict between two features	If the pin is made easy to lock and open, wear gets worse
Inherent contradiction	The clearance between the pin and the part should be small, *and* the clearance should be big
Intensified inherent contradiction (if it can be intensified)	There should be *no* clearance between the pin and the part, *and* there should be big clearance

4.4 SUMMARY

Numerous problems and tradeoffs can be boiled down to a single inherent contradiction. The contradiction is intensified as much as possible. The crazier the contradiction, the better the solution. The solution of the inherent contradiction removes many problems and gets many benefits at one stroke.

Terms like contradictions, paradoxes and conflicts are now increasingly used across industries and in many business areas.

Table 4.7   A Template for the Study of Your Own Problems

Modeling steps	Your Example
Visible drawback	
Tradeoff: the conflict between two features	
Inherent contradiction	
Intensified inherent contradiction (if it can be intensified)	

REFERENCES

1. Peters, T.J. and Waterman, R.H., *In Search of Excellence*, Harper & Row, New York, 1982, 91.
2. Field, S., *The Screen-Writer's Workbook*, Dell, New York, 1984, 31.
3. Altshuller, G.S., *And Suddenly the Inventor Appeared*, Technical Innovation Center, Worcester, MA, 1996, 21.
4. Savransky, S.D., *Engineering of Creativity*, CRC, Boca Raton, 2000, 235.
5. Altshuller, G.S. and Shapiro R.B., Psychology of inventive creativity, *Izobretenie, II*, 23, 2000.

5

MAPPING OF INVISIBLE RESERVES

INTRODUCTION

In Chapter 2, we defined resources as things that are available but are not being used — sometimes we can't "see" them because we have too many biases. In Andersen's fairytale, a little child saw that the emperor had no clothes. The ancient Mayans used wheels for toys and obviously knew how to make wheeled vehicles, but they never built them for any other uses. We recommend you review the examples in Section 1.1 before proceeding with this chapter.

Our use of the word "resource" in this book is nearly the same as in common language. Our list of resources includes materials and energy, human resources, information resources, etc. The word "nearly" indicates an important limitation: we are interested in idle resources available in the system and its environment, not in all resources. We are interested in free or very cheap resources, not in expensive additions.

First, we consider the invisible reserves of systems. All systems have gray zones or proximal zones of development. They are areas where we can find solutions that have potential to be developed, but have not yet been developed. They are zones where business opportunities lie.

Second, we look at the benefits from resource analysis. Understanding resources will help you in many ways. The analysis of resources in a situation can independently stimulate new ideas. Understanding resources can resolve the inherent contradiction that creates a problem in the first place. The analysis of resources will help you to foresee the evolution of the system and to understand customer needs that you have not previously identified.

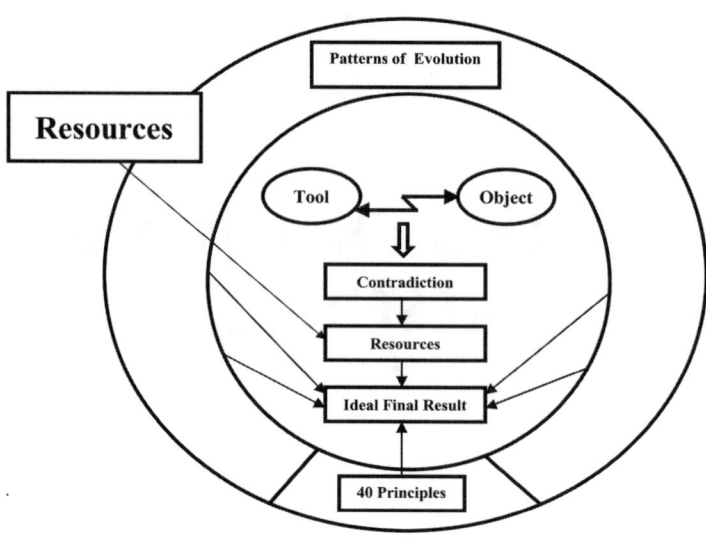

Figure 5.1 Consider all the resources. The mapping of resources can be used as both a step after defining the contradiction and as an independent tool.

Third, we present a simple, handy classification of most useful resources. The tool, object, environment and macro- and microlevel systems are resources classified by their relationship to the system. On each of the system levels, a variety of resources may be available: substances, energy, space, time, etc. Other resources, such as information, people's skills and solutions used by other industries are also available.

Fourth, we study the seven most important resource groups in detail. The fifth part outlines using resource analysis for explaining undesirable phenomena, for example, occasional faults in products with no visible reason.

In the model of problem solving (see Figure 5.1) resources work as a bridge between the contradiction and the ideal final result. As we have already said, resource mapping can also be used as an independent tool.

5.1 INVISIBLE RESERVES

Frequently, everything needed to resolve the contradiction is available, but the conflict has not yet been solved. Resources lie in a zone between the current level of technology and a more ideal, but feasible level. The relationships among contradiction, resources and the ideal final result are shown in Figure 5.2.

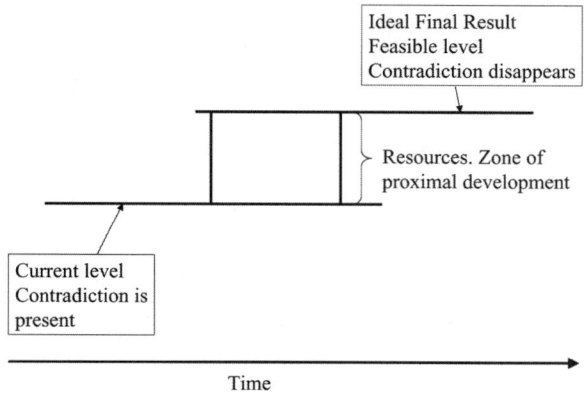

Figure 5.2 The relationship among contradictions, resources and the ideal final result. In time, new resources appear. After a time lag that sometimes can be very long, they are found and used. One benefit from resource mapping is reducing the time lag.

An analogy from a person's development may help explain the use of resources. People cannot learn everything, but they are capable of learning some skills in given time. For example, an average person can learn a foreign language over a period of time, but not 20 languages. The skills that people do not yet have but can develop for themselves are called the zone of proximal development. Correspondingly, we can say that resources define the zone of proximal development in technology — solutions that can be developed but have not yet been developed.

In the evolution of the lawnmower, one can easily see the three concepts we have spoken of:

Current level of the conventional lawnmower with contradictions. The device cuts grass well, but is noisy. We want a big muffler, but we want no muffler. Grass and a small duct directing air flow are available, but are unused resources. The ideal final result: grass is used as a muffler.

The carrot seeding case also illustrate invisible resources rather well:

1. Current level of seeding by hand. There are conflicts: many seeds–one seed, high speed–low speed.
2. Biodegradable materials for fixing seed spacing, such as paper and straw, have been available for a long time.
3. Ideal final result: tape for positioning seeds.

The latch mechanism case:

1. Traditional latching mechanism with contradiction: big clearance–no clearance.
2. Geometry has always been an available resource.
3. The change of geometry, that is, changing the form of the pin from cylindrical to conical, is used.

Figure 5.3 presents one of the cases, carrot seeding, in visual form. Figure 5.4 is a template for presenting your own examples.

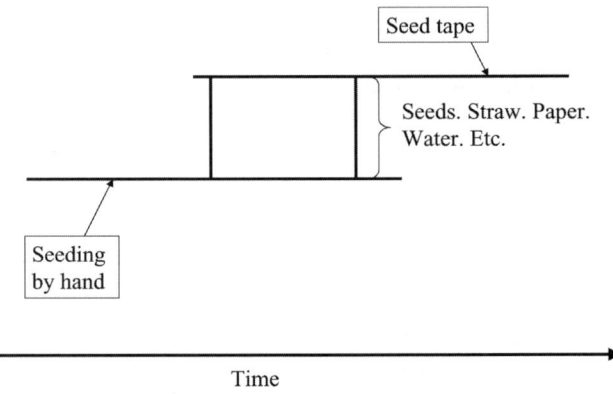

Figure 5.3   The evolution of the carrot seeding.

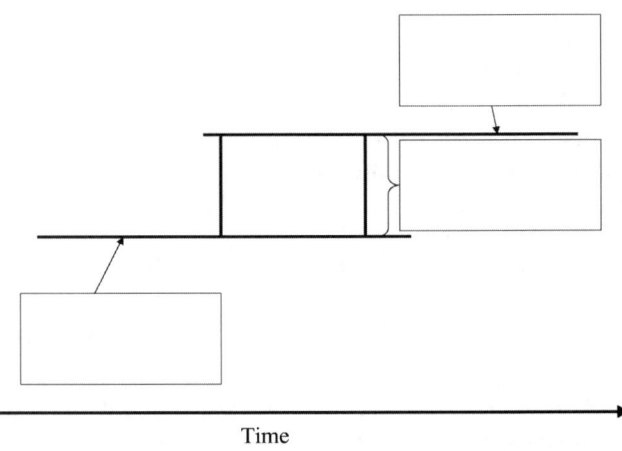

Figure 5.4   Make your own examples of how resources appear and will be used.

5.2 USING THE CONCEPT OF RESOURCES

Resource analysis is a handy tool with many benefits:

- Getting new ideas directly
- Solving contradictions
- Predicting the system evolution

5.2.1 Getting Ideas

Mapping resources stimulates ideas about how to improve the system. For example, sometimes it is enough to write down resources such as "empty space" and "geometric form," and new ideas appear. Hanging bicycles from the ceiling to be able to store more items in a garage is an example for use of "empty space." Recognizing that the flat bottom of a pizza box is the principal cause of heat loss, so that making the surface corrugated can solve the problem, is an example of using "geometric form."

In many business problems, the customers or users themselves are a resource. Many new types of self-service, including self-education, self-diagnosis (with home test kits for medical problems), self-treatment, self-planning (of one's house, for example), are emerging as businesses that recognize the customer as a resource.

5.2.2 Solving Contradictions

If you have formulated the inherent conflicts that we considered in the last chapter, you can use resources to resolve them. Chances to get good ideas increase if you have a complete list of the resources available to solve the problem.

5.2.3 Forecasting the Evolution of Technology

When we know the available resources, we know some features of the near future. Somebody, somewhere, sometime, will inevitably put those resources to use.

5.3 DIFFERENT RESOURCES

Resources abound. Actually, we can make an endless list of resources if we examine the problem from greater and greater distance. How, then, can we select the most useful resources? The following grouping of primary resources is one helpful technique. Groups are also illustrated in Figure 5.5.

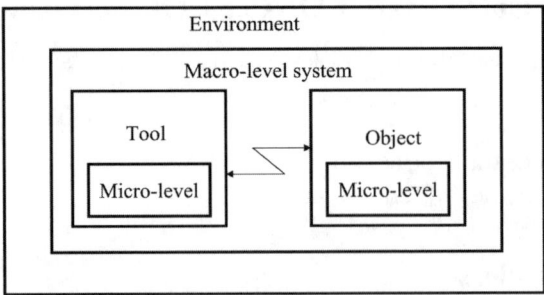

Figure 5.5 Mapping resources. List the resources found in the environment, the macrolevel system, the tool and the object. The tool and object can contain microlevel resources and their geometry; features and properties of the materials they are made from are resources, too.

First, divide the resources by system levels:

- Tool
- Object
- Environment
- Macrolevel system
- Microlevel system

On all system levels, a variety of resources are available:

- Substances and things
- Modified substances and things
- Voids
- Interactions and the energy to make them happen
- Form
- Features or properties
- Space
- Time

Other resources include:

- Information
- Harmful substances and interactions
- People's skills and abilities
- Etc.

You will find it helpful to examine your problem, listing everything that falls into each category. This can be called the TRIZ "treasure hunt" to find every treasure that can be used to solve the problem. This classification is a generic template, not a set of rigid rules — use it to help you start your treasure hunt. More detailed classification can be found in the literature on TRIZ, for example, in Savransky's book.[1]

Let's use the examples from the previous chapters to illustrate each of the categories of resources.

5.4 THE MOST IMPORTANT RESOURCE GROUPS

5.4.1 Resources of the Tool and the Object

The case of the latching mechanism is a typical example of the tool itself as a principal resource. In the lawnmower example, the muffler could be changed to some extent, but the most creative solution was to use a component from the macrosystem — grass. In the case of cultivating carrots, the tool was the human hand, which is difficult to change. No wonder that in this case, the object (the seed) was used as a resource, but was modified by the addition of the tape, a resource that is easily available in the environment.

The tool is often, but not always, an excellent choice as the resource to use or to modify:

1. The efficiency of fighting fires has been improved by changing the tool — water. First, water was poured on fires using buckets, then pumped through hoses, then sprayed in droplets using modified hoses and, at last, delivered as an atomized mist.
2. The performance of a cutting edge has been improved by changing the materials and the geometry of the edge, the geometry and the materials and the coating of teeth (on the edge of a saw), then by replacing the blade by powder, liquid, plasma and laser. The function of the tool is still to cut the object, but the changes have been made to cut different materials, to increase controllability, to increase speed, etc. The history of cutting tools illustrates creative use of resources to respond to changing customer needs.

In business problems such as training and marketing, many tools can usually be changed. They include plans and curriculum, training and presentation materials, skills of personnel, etc. So, check the resources of the tool first.

Collect your own examples of how the tool is used as a resource (see Table 5.1).

The case of cultivating carrots is a typical example of the use of the object's resources. Many seeds are combined and can be handled as one big "seed." The same concept is used when packages are combined for transportation.

Table 5.1   List Three Examples Where the Tool was used as a Resource to Solve a Problem

Problem	Tool	Object	Solution

Many nontechnical systems can be improved using the resources of the target group served. Fast-food restaurants, supermarkets, department stores, information Web sites and innumerable other businesses that rely on self-service are perhaps the most common examples. One of the most effective ways of marketing is to deliver products and services that are so good that happy customers advertise them to potential new customers by word of mouth. Buyers themselves act as salespeople.

Collect some examples of the object as a resource, too (see Table 5.2).

The object cannot, unfortunately, always be used as a resource. But always check.

5.4.2   Resources of the Environment

Environmental resources mean things and the substances of which they are composed, energy and the fields that always surround us. They are often ignored, because we see them every day. These resources include air, water, empty space, gravity, sunlight and other free natural resources. They can also include the resources that are available in a specific situation. For example, most factories have people, compressed air, electricity, information networks (sometimes computer based, sometimes paper based), hot and cold water, etc., available. Most office buildings have windows, telephone and data systems, electricity, light, people, etc. Hospitals have medicines, doctors, nurses, technicians, patients, beds, imaging equipment, information systems, water, air, light, etc. It is easy to see how something might be a tool in one situation, an object in another and part of the environment in a third.

In northern countries in past times, ice was collected in winter, stored under a layer of sawdust and used to cool milk and other food products in the summer. This technology became outdated when refrigerators were introduced, but has been resurrected in another form. In the Swedish city of Sundsvall, snow and ice are collected in winter and stored in a pit insulated with wood chips. The runoff is used to cool offices in summer.

Table 5.2 List Three Examples Where the Object was used as aResource to Solve a Problem

Problem	Tool	Object	Solution

If cold can be used as a resource, why not excess heat? In some places in central Europe, heat is collected from roads in heat accumulators and then used to keep the same roads dry in cold weather. Likewise, heat pumps use the heat available in the environment as an energy resource to operate cooling systems.

Solar energy is, of course, an environmental resource that is being widely used for many purposes. Solar-powered space satellites, solar-heated water and solar-powered lawnmowers, for example, have all been available for years and span quite a range of technical complexity.

Empty space is one important invisible resource. A garbage bin partly hidden underground uses space beneath. Empty space is sometimes used in firefighting instead of water and other substances. Firefighters attack forest fires by building fire "breaks" or clearings where there is nothing to burn. In a hot kitchen stove, food in the oven can sometimes catch fire when the oven door is opened. The best way to extinguish the fire is to close the door calmly. When the oxygen is consumed, the fire will go out.

When solving business problems, sometimes the best thing to do is nothing. Organizations often try to improve results by emphasizing team-work and holding many kinds of meetings. However, results often improve if the number of meetings is decreased. Many people complain that they have no time to do anything, but they may solve problems making decisions not to do something. For example, we all know people who feel they have a moral responsibility to answer all e-mail messages immediately. An obvious solution is to throw away most messages and to answer some messages later. Time management and personal organization are popular topics for seminars and consulting. For many years, people have systematically excluded useless operations in the production of cars, computers, washing machines and other tangible products, but there is still much "non-value added" work (to use a popular buzzword in the process analysis world) in the processing of information and the management of organizations.

5.4.3   Using the Macrolevel

Any system can be combined with other systems into one greater system. Both similar and different systems can be combined. Practically everything can be improved in some respect by transition to the macrolevel. Sometimes the effect can be dramatic. The frame and cover in a riding mower produced by the John Deere Company at one time consisted of 153 steel parts. When plastics replaced metal, only three parts were needed. The other 150 parts were integrated into the higher-level system.[2]

In the lawnmower case, one can try to resolve the conflict between big muffler–small muffler using resources on the higher system level. Grass is only one of the resources that can be used as a muffler. There are others. The casing of the lawnmower can work as a muffler. Lawnmowers are often equipped with bags for collecting grass. Obviously, the bag could be used as a muffler, too. Environmental laws are being strengthened in many regions and catalytic converters may be required for small engines. If the converter is necessary anyway, why not try to get additional benefits by using it as a muffler, too?

What happens if we combine grass and lawn? Or if we consider many lawns instead of one? What could that mean? The whole job could be "outsourced" to a service company that works on many lawns. It can use quiet electric machines that are often too expensive for small lawns. An analogy is a central vacuum cleaning system in an apartment house. Instead of a vacuum cleaner in every apartment, there is one cleaner for a whole building and only nozzles and tubes in each flat.

Seeds were combined into a larger system: tape with seeds. We can continue combining at higher and higher levels. What happens if we add fertilizers, soil, water, air and other substances necessary for growing carrots to the tape? Lior Hassel, an agricultural engineer in Haifa, Israel, has developed a system that grows vegetables inside standard metal shipping containers using hydroponics. A robot-controlled system is producing 500 heads of lettuce per day, a yield 1000 times greater than a similar area can produce using conventional farming. If lettuce can be produced this way, why not carrots? Surely there are many intermediate solutions between a tape with seeds and totally automatic farming. Revolutionary changes in gardening and farming are on the way in all parts of the world.

If you have a solution on some system level, look at higher and lower levels. Sometimes you will find unexpected new solutions. In the simple latch pin problem, gates, hydraulic and electric actuators and other auxiliary parts were changed.

In marketing and training, macrolevel resources are used even more than in technological systems. The same advertisement or the same subject matter in training is repeated many times. This is a simple combination of similar objects; modern learning theory suggests that the course should present each idea in every learning style, so that all students can benefit. Advertisers learned long ago to deliver the same message in a wide variety of channels — everything from audio to video to painting the message on a racecar or a passenger bus. Entrepreneurs are building networks all the time to provide more channels. We can view networks as resources.

5.4.4 The Microlevel Resources

The opposite of moving to the super system or the macrolevel, is to go "down" to the microlevel. The system is segmented into smaller parts. In our example of the evolution of firefighting with water, the example of reducing water to smaller and smaller parts also illustrates the transition to microlevel. The conflict "much water–no water" is solved using microlevel parts.

Microlevel parts can be empty. Pores, capillaries, holes and the use of space are frequently called "use of voids" in TRIZ. Remember, a void is not "just nothing" — the void may have structure or texture that gives it a function to solve your problem. Textiles for outdoor clothing should be impermeable to rain and at the same time permeable to water from perspiration. Gore-Tex and other materials with micropores resolved the contradiction. The pores allow water molecules to pass through, but stop water drops. Foams and gels are highly technical ways of using voids (capturing "nothing" and making it do something useful.) This is technical, but it isn't new — whipped cream uses the chemistry of the fat molecules to capture the air and create a new product.

Segmentation may also mean that many small machines replace a big one. Imagine that, instead of a conventional lawnmower, many automatic minimowers resembling turtles are working on the grass. They may easily work on lawns with complex shapes. Distributed computing is a popular system — both within companies and on public projects such as SETI (SETI is an experiment that uses Internet-connected computers in the Search for Extraterrestrial Intelligence). The resources of many small computers on a network are used as elements of a larger computer whenever they are idle.

Let's consider the case of the latching mechanism. One can try to resolve the conflict "big clearance–small clearance" using different resources on the microlevel. What does this mean and how can one get there? The simplest way is to segment the pin; divide it into two parts. One of the parts may be a wedge that ensures tightness when the latch is closed. Segmentation can be continued. There may be many parts, either layers or filaments. If the segmentation is continued, more ideas appear: a pin made of powder, gel, liquid, gas and their mixtures. Would you consider a dynamic pin that changes its size using some microlevel effect?

Many examples of segmentation can be presented from the areas of communication, business and education. One perpetual problem in communication is the writer's problem of sending and the reader's problem of receiving (quickly, easily and accurately.) a message that has rich content. Charles Dickens' novel *The Pickwick Papers* was first published in the 19th century in 20 monthly parts in a journal. It was the first serialized novel and a literary forerunner of radio and TV series in the 20th century. All Dickens' novels were first published in parts and had enormous success. Readers found parts easier to read. Further, they got an extra benefit: a new dimension of suspense.

5.4.5   Time Resources

Problems and contradictions can often be solved using different properties of the system at different times or by modifying the system so that it has different characteristics at different times. In carrot cultivation, the thinning operation should be and should not be. The solution realizes these requirements that, at first glance, seem incompatible. The row of plants is not thinned after the plants germinate, but it is thinned before planting.

The conical pin is tight when working and loose in the sense that is easy to insert and remove.

Biodegradable screws to hold broken bones in alignment vanish when they are not needed anymore. The operation for removing the screw, too, vanishes from the surgical system.

Modular systems are often used to make it easy to modify the properties of a system over time. A modular bookcase can easily be made smaller or larger when the owner's needs change.

In business problems, we can seek different time reserves. Sometimes we have too little time and sometimes we have too much. Nearly always, some time resources can be found somewhere. Examples are given in Section 5.4.2.

5.4.6   Space Resources

The conical pin separates properties not only in time, but in space. It is loose or easily movable, in the direction of its axis and tight in the perpendicular direction.

In the lawnmower case, one can try to resolve the big muffler–small muffler conflict using some space available. See earlier examples using the case of the lawnmower or the bag for grass as a muffler.

In the first chapter, we used examples of the garbage bin and ax. Both were improved using space as a resource — space beneath the bin and space inside the handle.

How to use space resources in business? The resources here are different places that can be used for work and study: home, train, bus and airplane. In conventional business, the customer is in one place and the supplier is in another. In many emerging business models, the customer is brought inside the business, either virtually (e-business systems, shared information) or physically (customers participating in product design, suppliers building specialized minifactories inside their customers' facilities, specialty boutiques inside department stores, etc.).

5.4.7 Other Resources

There are other important resources. All products have some esthetic appearance and they provide information. In the ideal case, the system works well, looks good and tells how to use it. For example, a well-designed door tells you how to open it without labels like "push" or "pull."

Harmful substances and interactions can be considered resources too, sometimes called blessings in disguise. For example, wastes (materials and energy) that can be used in another place or recycled are resources, too. Many medicines have been developed from poisons, by finding ways to protect the healthy tissue while destroying the unhealthy parts.

Materials and technologies known in other industries can often be used to solve problems. They can also be considered resources. Microwave ovens came from radar technology. Cutting with water, Gore-Tex, remote cardiac measurement and many other technologies and materials were the results of space research. Production technologies and tolerances of machine building are slowly spreading into the construction industry. Networking methods are spreading from business into education.

Anything that anybody else has done to accomplish a particular function is a resource for solving a problem. The better the definition of the problem, the easier it will be to find those resources. By expressing the problem in nontechnical language, you may help yourself find the resources developed in other industries. For example, a farm group was trying to find a way to dry cow manure without using conventional heating methods. But the word "dry" kept leading them to heat. When they redefined the problem as separating liquid from solids and searched a variety of databases, they found a technology using hydrophilic molecules to carry water away from a liquid mixture.

This method had been used for 40 years to concentrate fruit juices without heat.

Technologies transferred from other industries need not be exotic. Mitchell Weiss, age 13 years, has used the bicycle to invent a pedal-powered lawnmower: "You work it like a regular bicycle. You pedal it to turn the back wheel, which pushes the front lawnmower wheels, which are attached to the blade."[3]

5.5 WHEN RESOURCES ARE IN USE BUT SHOULD BE REDISCOVERED

Sometimes we don't need a new useful action but we need to explain why and how harmful action develops.

At the time of Prohibition, smugglers developed a lot of smart solutions to avoid the police, many of which have become legends. One is the smuggling of liquor by boat. Inspection of the boats never found any bootleg booze. Everybody knew that it was smuggled by sea. Bribes were not an explanation — the police were honest. What was happening?

The smugglers had a contradiction: liquor should be in the boat to run their business and should not be in the boat to avoid problems with police. The first part of the solution was to use time resources. It was enough to get rid of the liquor just when the inspectors were on board. One obvious solution was to sink the containers underwater, combining them with some heavy thing. But how to get them up again? The real, inherent contradiction was that the containers should be heavier than water to disappear when needed and lighter than water to appear when needed. The extra weight should itself disappear. What could be the weight that disappears easily in the water? Something that dissolves in water? Which substances are easily available and dissolve in water? Sugar? Salt? Smugglers used table salt as a weight that dissolved and the containers surfaced at a planned time (see Figure 5.6). Of course, the police eventually learned the trick and today's drug smugglers have to use different methods.

5.6 SUMMARY

■ Seek idle, invisible, free or very cheap resources in the system and its environment.
■ Use all the benefits from resource analysis: getting ideas, getting ideas with other tools together, forecasting the evolution of systems, forecasting customer needs.

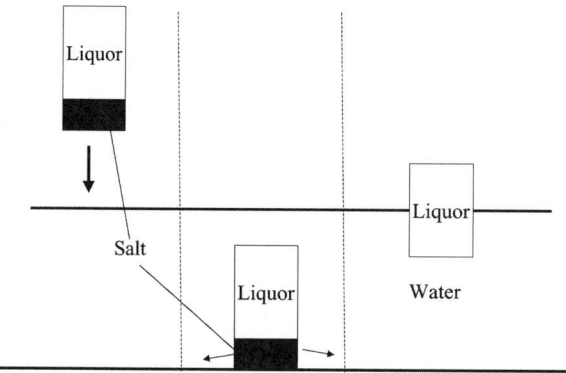

Figure 5.6   Salt and water are resources that made liquor containers first disappear and then appear again.

- Map many resources, not only those that are most obvious. A deep analysis pays back. Often, unexpected opportunities can be found.
- A good checklist for resource hunting is the following: tool and object, environment, high-level system, microlevel, time, space, etc.

In this chapter, we have considered the resource analysis mainly as an independent tool. In the following chapter, we will study in detail how resources are used to solve the inherent contradiction and to define the features of the ideal final result. We will tell how to select the principal, primary, most important resource and how to use resources together.

REFERENCES

1. Savransky, S.D., *Engineering of Creativity*, CRC, Boca Raton, 2000, 83.
2. Smith W.E., *Principles of Material Science and Engineering*, McGraw-Hill, New York, 1996, plate 2.
3. *Popular Science*, October 2001, 31.

6

THE IMPOSSIBLE OFTEN IS POSSIBLE: HOW TO INCREASE THE IDEALITY OF THE SYSTEM

INTRODUCTION

In Chapter 2, we identified the ideal final result as the solution resolving the contradiction. In this chapter, we will study the ideality of the system in detail.

We will first study the ideal final result as an independent tool. We also briefly examine different ways to describe ideal systems.

Second, we will study how to go from the definition of contradictions and resources to the ideal final result. In previous chapters, we defined good solutions as those that achieve the ideal final result and resolve the contradiction using idle resources. This is easy to say but hard to do. To make it easier, we need a systematic method for using resources to remove contradictions. For this, we introduce the concept of the principal resource. In Chapter 3, we described numerous different tradeoffs. In Chapter 4, we showed how to find the inherent contradiction behind the bundle of tradeoffs. In Chapter 5, we made long lists of resources from inside and outside the system. Now we will select a single, primary, most important resource, called the principal resource. Other resources are auxiliary. They help the principal resource remove the inherent contradiction. In this chapter, you will learn to construct the ideal final result from principal and auxiliary resources. In Figure 6.1, we review the general model for problem solving.

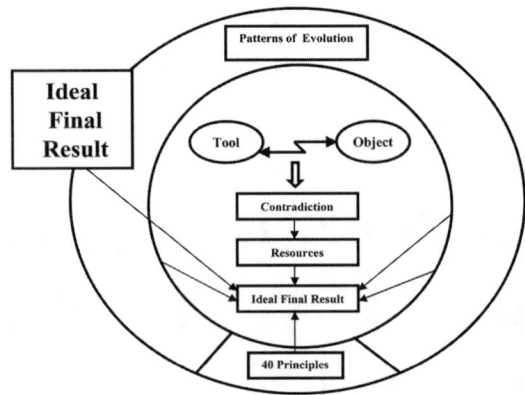

Figure 6.1 The ideal final result in the model for problem solving. The arrows show that the ideal final result can be developed by the application of many different tools.

6.1 THE LAW OF INCREASING IDEALITY

The concept of the ideal final result is based on the law that Altshuller first formulated as follows: "The development of all systems proceeds in the direction of increasing the degree of idealness."[1]

In other words: systems get simpler, not more complex. The ideal system is the one that has all the useful features and functions of the original system, but has no weight, no volume, requires no labor, no maintenance, consumes no energy, etc. Where does this definition come from? Start with the ideality equation:

$$\text{Ideality} = \Sigma \text{ Benefits}/(\Sigma \text{ Costs} + \Sigma \text{ Harm})$$

The Greek symbol Σ means "the sum of," so this equation reads, "Ideality is the sum of all benefits divided by the sum of all costs and all harm." If you dislike formulas and equations, don't worry. The formula is qualitative. We will not do any actual calculations. The point of the formula is that it clearly illustrates two sides of ideality.

The formula generalizes numerous expressions presented to describe the level of technologies, inventions and solutions. It was adapted from the value equation of *Techniques of Value Analysis and Engineering* in the early 1950s:[2] Value is the capability of the function divided by costs ($V = F/C$). Benefits can contain functional capabilities, but are not limited to them. Many important features, such as weight and size, are not actions or functions.

In his book *Great Inventions through History*, Gerald Messadié compares inventions with the fishing technique of the seagull: "The seagull, which carries a clam in its beak, places it on a wall, then goes to pick up the biggest stone it can manage and drops it from a height in order to break the shell, has invented the technique which is inspired by neither the spirit of commerce nor the desire for power. This technique enables it to obtain its food reasonably quickly and in return for a little ingenuity .... Thus, it saves time and consequently energy: this is the goal of absolutely all the inventions which have been made since humanity began .... All inventions are included in this absolute rule."[3]

The rule formulated by Messadié stresses very important features, but is nevertheless too narrow: The increase of ideality could be a faster, more energy-efficient system, but lots of other changes are possible, too: decreasing weight and size, improving outer appearance, increasing comfort or decreasing harmful byproducts, for example. Nevertheless, Messadié's rule is not a bad illustration of the law of increasing ideality. Examples can be found easily. Observation of our world shows that time or energy or both are saved in many improving systems, from those as simple as our carrot-planting example to the development of computers and transportation systems. Cars and airplanes are means of traveling more quickly and with the least possible effort. Electric lighting is extremely easy to use compared with torches and earlier lamps of any kind. The same can be said of food processing, communication technology and all other technologies that are widely used — their commercial success is proof that they have made life easier for people.

The maximum value of the equation is reached when the denominator is zero — that's how we concluded that the ideal final result is a system that achieves all the benefits with no cost and no harm. Because complexity causes increase in cost and harm, increasing simplicity will increase ideality if the benefits stay the same.

Ideality is measured by comparing systems. We can easily say which of two alternative systems is closer to ideal in specific circumstances by comparing the ideality equation. If benefits increase with no change in cost or harm, then ideality increases. If harm is reduced with no change in benefits or cost, then ideality increases. Some examples:

■ In the 1970s, the pocket electronic calculator quickly replaced the slide rule. The calculator was more precise, easier to carry and easier to remember how to use. When the price of the calculator fell, the slide rule quickly became a museum piece.

■ Several rotating piston engines for cars have been developed in an attempt to find an alternative for the conventional internal combustion engine. Ideas for replacing the linear motion of pistons with rotary motion were developed early in the history of the steam engine — James Watt got a patent for a rotating steam engine in 1769. Rotating machines have their appeal. Steam and gas turbines are good examples. Nevertheless, a rotating engine has not superseded conventional ones in cars. The benefits are too small compared with the drawbacks, such as sealing problems and the high consumption of fuel. A clear increase in ideality and a clear supremacy over the competing system are absent.

■ Scissors formed to fit the human hand (Fiskars and some other brands) have become popular and partially replaced conventional models. A small change of geometry produced a noticeable improvement in comfort.

■ Telephones and many other products today have much simpler, less complex forms than they did early in their development.

■ Corporate logos and fonts in typography have become simpler, too. Although fashion may be at work here as well as the evolution of systems, the same message is delivered using simpler and leaner forms.

■ Control of automobile braking and steering systems "by wire" instead of by direct mechanical linkage, is 20 years behind the development of "by wire" systems for aircraft. But, it is advancing quickly and in the direction of increasing ideality. The new system can stop a car in an emergency in 10% less distance than conventional systems, using 15 parts instead of 45.[4]

Sometimes scholars, who view technology as one part of the social organism, see important features better than engineers. Historian Arnold J. Toynbee gives some excellent illustrations from the evolution of technology and science, where he sees "a law of progressive simplification." First, transportation:

"...When the horse was replaced by the locomotive, the simple carriage-road had to be turned into an elaborate 'permanent way,' with ... a pair of metal rails ... in the next stage of technical advance, when the ponderous and bulky steam engine ... is replaced by the light and handy internal-combustion engine, ... the improvement in technique is accompanied by a notable simplification of apparatus .... The technical advantage of mechanical traction is not only preserved but enhanced (inasmuch as the internal-combustion engine is an improvement on the steam engine from the mechanical standpoint); and at the same time the disadvantage of the elaborate material apparatus is partly transcended. For the motorcar liberates itself from the rails to which the locomotive is bound and takes to the road again, with all the speed and power of a railway train and almost all the freedom of action of a pedestrian or a horse."[5]

Toynbee also has illustrative examples of the evolution of telecommunication, writing, fashion and astronomy. The telegraph and telephone were invented first and transformed business and society with a speed of communication not imaginable earlier. More recently, wireless communication has dramatically improved the transmission and accessibility of information.

Fashion has evolved from the extravagant costumes of Queen Elizabeth or King Louis XIV toward plainer materials and simpler cut. The triumphal march of denim and "casual" clothes even in traditional businesses has continued the trend in this century.

Astronomy and physics have long recognized the usefulness of increasingly simple models and theories. The Ptolemaic system (earth-centered) had to postulate complex epicycles to explain observed movements of known heavenly bodies. The Copernican (sun-centered) system presents in far simpler terms, wider range of movement of innumerable bodies. Modern theories of elementary particles use six (or so) quarks to describe all other particles — in the 1960s, more than 200 particles were used to describe matter.

Let's consider some typical ways to improve systems. If the useful features are clearly much improved and the greater numbers of parts cause very little harm or cost, increasing the number of parts can improve ideality. A modern bicycle has more parts than the first "hobby horses" in the late 18th century. The features are improved so much, however, that the ideality of the bicycle is increased. Overall, over history, the number of parts will decrease. The new electronic gearshift mechanisms are much simpler than the mechanical shifting systems that have dominated bicycle technology for the last 50 years.

Another way to increase the ideality of a system is to decrease the size, energy consumption, weight, the number of parts and operations and other cost-generating factors. The frame of the bicycle is usually made of a few parts. A single-part frame is used for sport bikes. The ideality is improved by decreasing the number of parts. Or course, ideality was improved based on a specified goal. If the goal is different, the ideality may or may not, be improved.

Instead of the term "ideal system" we can use terms like "ideal machine," "ideal process" and "ideal substance" (or "ideal material"). The ideal machine does not exist, but its job gets done. The ideal process is one that consumes no energy and no time, but produces the desired product or service. The ideal substance is one (nonexistent) having all needed features. Here are examples of each variation:

- I want to produce parts, but I don't want to build a factory with lots of specialized machine tools. The current solution, which is on the way to an ideal solution, is the stereo lithography system that can produce parts from a three-dimensional CAD file and a selection of metal, plastic and ceramic powders.
- I want to go to exotic places, but I don't want to spend time traveling. Old solution: read a book. New solution: visit Web sites that have real-time cameras in exotic places. The *Star-Trek* solution — instantaneous travel — is impossible now, but may be possible in the future.
- I want clean clothes, but I don't want to do the work of washing them. Two pathways exist up this "mountain." Along one path are services that take your clothes, wash them and return them without any work on your part (but the cost is not zero). Along the other path are machines that wash and dry clothes with increasing sophistication (sorting, selecting the right temperatures and wash methods, etc.) but the machines and the chemicals also have non-zero cost. The next step on this path, currently being tested by several manufacturers, is the incorporation of ozone-generating materials in the machine so that soap or detergent will not be needed, reducing the operating cost and reducing the harm that the wastewater does to the environment.

In the ideal system, the harmful feature disappears and the useful one is retained. The solution gets closer and closer to the ideal if the harm can be turned into a benefit or a "blessing in disguise," as scrap merchants have traditionally turned waste to profit. Later design for remanufacture, design for recycle and, generally, design for the environment illustrate increasing ideality. Examples:

- Biological waste can often be converted to valuable and environmentally friendly bio-gas.
- Wastepaper becomes raw material for recycled paper and packaging products. For this reason, experts joke that the biggest producer of paper in the world is the city of New York.
- Customer complaints are certainly undesirable, but if a company learns from the complaints and creates a better product, then the harm turns into good.

Add your own examples to this list.

6.2 CONSTRUCTING SOLUTIONS FROM RESOURCES

Sometimes the formulation of contradictions and the mapping of resources tell us nearly directly how to solve the problem. If we know that the garbage bin should be small *and* large and that geometric space is one of the resources, we can rather easily discover the idea of using the space beneath the bin. The visible bin remains small and the bin as a whole gets large.

The information or resources may, however, not be enough to find an idea for the solution. In Chapter 5, we considered the smuggling problem. We know that the canisters should disappear at certain times and appear again at another time. Analyzing contradictory requirements for the tool, we concluded that the canister should be heavier *and* lighter than water. Then the canister can be dropped temporarily into the water and sink under the surface, but later float at the surface for easy recovery. Resources are: a canister, liquor, water, buoyancy of water, gravitation and resources of macrolevel systems. This information may, however, not be enough to find an idea for the solution.

Additional steps are needed to move from the resource analysis to the ideal final result. First, we select the principal resource. Remember that the principal resource is the primary, most important resource or the resource that exhibits the inherent contradiction. If you have defined the inherent contradiction, described in Chapter 4, it is rather easy to find the principal resource. Recall some examples:

- In the smuggling problem, the canister should be heavy *and* light. The principal resource is the canister.
- In the lawnmower problem, the inherent contradiction is big muffler-no muffler. The principal resource is the muffler.
- In the carrot-cultivating problem, the inherent contradiction is many seeds-one seed. The principal resource is a seed.
- The problem of the latching mechanism: the contradiction is no clearance-big clearance. The principal resource is the pin.
- In the training problem, lots of time is needed for training and no time at all is available. Time is the principal resource.
- In the firefighting example, we need much water and no water. Water is the principal resource.

Auxiliary resources can change the principal resource so that the contradiction disappears. The smugglers found an excellent auxiliary resource: salt, making the canister first heavy and then light again. In the lawnmower problem, grass helps make the muffler smaller. In the carrot-cultivation problem, something connecting seeds helps make "one seed" from many seeds. In the example of the latching mechanism, the geometry of the pin itself makes the clearance change from big to zero. In the training example, working time is the resource that can be used to minimize training time. In the example of firefighting, high pressure makes much water from almost none (mist).

Table 6.1   List Examples Illustrating Increasing Ideality in Systems with which You are Familiar

Initial System	
Improved System	
What Changed?	
Benefits Improved	
Cost Reduced	
Harm Removed	

These examples are simplified, of course. Some thinking is needed to find proper auxiliary resources. Smugglers, obviously, used some time to figure out that there is salt available. It is useful to list more than one auxiliary resource. Sometimes, changes of the principal resource are needed to get a good solution.

To review the formulation of the ideal final result using resources requires three steps:

1. Select the most important or primary resource having an inherent contradiction. See Chapter 4 on intensifying contradictions for help.
2. List auxiliary resources or resources that can change the primary resource. See Chapter 5 on resources.
3. Change the principal resource by using auxiliary resources so that the contradiction vanishes.

These steps can be conveniently organized in a table (Table 6.2) and illustrated using our examples. Many auxiliary resources will make other solutions possible.

In the noise problem, an exhaust tube, exhaust gases and grass make the big muffler small or, even better, make a big muffler into an "absent" muffler (Table 6.3).

If we use different resources, we'll get different solutions. For example, if the casing of the lawnmower is used as an auxiliary resource to redirect the sound and to absorb it, the case, instead of the grass, becomes the muffler.

In the example from carrot cultivation, seeds, soil and water and other resources change the seeds so that the number of objects is, in some sense, large and small at the same time (Table 6.4).

Table 6.2 Constructing the Ideal Final Result in the Smuggling Problem

Primary resource with the inherent contradiction: canister
High density–low density

Auxiliary resources:
Water, air, salt, sand, sugar, etc., time, gravity, buoyancy

Features of the ideal final result:
Salt changes the canister — heavy (high density) and then light (low density)

Table 6.3 Constructing the Ideal Final Result in the Lawnmower Problem

Primary resource with the inherent contradiction: muffler
Big muffler–no muffler

Auxiliary resources:
Grass, exhaust tube, exhaust gas, air

Features of the ideal final result:
Grass makes muffler present and absent at the same time

Table 6.4 Constructing the Ideal Final Result in the Carrot-Cultivation Problem

Primary resource with the inherent contradiction: seed
Many seeds–no seeds

Auxiliary resources:
Soil, water, waste paper (from food packages), straw, mulch

Features of the ideal final result:
Tape made from waste paper and other cheap materials combines many
seeds to "one seed"

In the example of the latching mechanism, the geometry of the pin is a resource that makes the clearance both large and small (Table 6.5).

In the training example, working time is used to get plenty of time when there is actually no time available. Work can be a form of on-site training if it is well designed, with lots of feedback so that the worker can learn from each experience. The longer employees work, the better trained they are. The solution is unconsciously used all the time.

Table 6.5 Constructing the Ideal Final Result in the Example of the Latching Mechanism

Primary resource with the inherent contradiction: pin
Big clearance–zero clearance

Auxiliary resources:
Pin: geometry, surface, material, time

Features of the ideal final result:
Geometry makes clearance wide when the latch is open and zero when it
is closed

Table 6.6 Constructing the Ideal Final Result in the Training Example

Primary resource with the inherent contradiction: time
Lots of time–no time at all

Auxiliary resources:
Working time, existing knowledge and skills, the culture of the company,
curriculum, textbooks, computer networks, students, experienced people,
teachers, etc.

Features of the ideal final result:
Work has a training effect. Training takes place over extended time periods,
without special training time, so that training gets better and work results
get better, too.

Consider one final example of firefighting (Table 6.7).

Examples help you to study your own system (see Table 6.8). Continue the exercise in the previous chapter and study how the resources you have mapped can be combined. You can also use the concept of ideality directly: decide what are the primary and auxiliary resources and how they can be used together.

6.3 SUMMARY

Study the difference between good and weak solutions. Increasing ideality is one important feature of good solutions. Numerous examples show that systems really can get simpler, even though they solve complex problems. Big benefits can be created with low cost and little harm. You can use the concept of increasing ideality directly: study good solutions in other industries and you will get ideas about improving your system.

Table 6.7   Constructing the Ideal Final Result in the Example of Firefighting

Primary resource with the inherent contradiction: water
Much water–no water

Auxiliary resources:
Water, water pressure, tubes, nozzles, air

Features of the ideal final result:
High pressure makes the amount of water nearly zero and very large–the volume of mist is very high.

Table 6.8   Study Your Own System

Primary resource with the inherent contradiction:
Auxiliary resources:
Features of the ideal final result:

If you have defined the inherent contradiction (Chapter 4) and mapped resources (Chapter 5), you can very effectively build the ideal final result from resources. Select the system with the inherent contradiction as the principal resource. Find auxiliary resources that change the principal resource so that the contradiction disappears.

The features of the ideal final result form the basis of the method for evaluating solutions. The evaluation and improvement of solutions are considered in Chapter 7. Different methods can increase the ideality of the system. These ways are called the patterns of evolution. The patterns are studied in Chapter 9.

REFERENCES

1. Altshuller, G.S., *Creativity as an Exact Science,* Gordon and Breach, New York, 1984, 228.
2. Miles, L.D., *Techniques of Value Analysis and Engineering,* McGraw-Hill, New York, 1961.
3. Messadié, G., *Great Inventions through History,* Chambers, St. Ives, 1991, 5.
4. *Forbes,* August 6, 2001.
5. Toynbee, A.J., *A Study of History,* Vol. III, Oxford University Press, London, 1963, 174.

7

HOW TO SEPARATE THE BEST FROM THE REST: A SIMPLE AND EFFECTIVE TOOL FOR EVALUATION OF SOLUTIONS

INTRODUCTION

Early in the book, we asked you to recall your best problem-solving experience and think about what characterized the good ideas. This chapter begins with another question. When you create a good idea, do you ever wonder: "Why not until now? Why didn't I think of this 2 years, 5 years, 10 years ago?" Companies tell us this story so often we have named it The Standard Story — *"A competitor introduced a new solution and we found the same idea in our own notes from many years ago."* Chapter 1 has many examples of good ideas that were neglected.

One of the most striking results from the authors' experience in teaching creativity classes and consulting on creativity is that recognizing, appreciating and evaluating solutions may be more difficult than finding them. Having good ideas is useless if they are rejected.

We hope you agree that it makes sense to seek better ways to evaluate solutions. In this chapter, we will present a simple and effective evaluation tool. We will study three points in this chapter:

1. We define the evaluation criteria, which we obtain from the concepts of ideality, contradiction and resources that we have studied in Chapters 3–6. Now we use these tools for a new purpose: evaluation of our proposed solutions.

2. We consider the measures of evaluation. The ideality of each proposed solution is evaluated and compared with the ideality of the other solutions. We use a simple and practical tool: pair-wise comparison with the known solutions. In real-life projects, the yardstick should be the best possible existing or developing competing methods or technologies. In the examples in this book, the solutions are usually compared with well-known current technologies for clarity and simplicity.

3. We discuss how to go further if the evaluation shows that we have not achieved the ideal final result. Sometimes, the whole idea may be bad and it deserves to be rejected. More often, the primary idea is excellent, but there are sub-problems that need to be solved. The evaluation criteria will help you see the path through the maze of problems and solutions and avoid confusion with numerous secondary tasks.

7.1 EVALUATION CRITERIA

When we have a new idea, we must ask, "Is this idea good or bad?" To answer the question, we need a set of criteria for good solutions. Let's first recall the ideality equation studied in the previous chapter:

$$\text{Ideality} = \Sigma \text{ Benefits}/(\Sigma \text{ Costs} + \Sigma \text{ Harm})$$

This is the basis for evaluation.

First, all harmful features disappear. Most often, problems are solved to remove some drawback, which is why it is logical to begin from this requirement.

Second, all useful features are retained and new benefits appear. We don't — and shouldn't — remove only drawbacks. New useful features should be introduced and existing ones retained.

Third, new harmful features do not appear. It is important to check this — a frequent problem with both business and product improvements is that the new system gets rid of the initial problem, but introduces more new problems. The software industry is legendary for "improvements" that cause customer dissatisfaction.

Fourth, the system does not get more complex (complexity increases cost and reduces reliability).

Fifth, the solution removes the inherent, primary, most important contradiction in the problem. Having studied tradeoffs and contradictions in Chapters 3 and 4, we can read this requirement in the ideality equation as well. To get benefits, we need more weight, size, energy, time and other cost-generating features. To cut cost and avoid harm, we should have less weight, size, energy, time or other properties, always ask what is the essential primary contradiction — and that is just what should be solved. Before the car was developed, vehicles used steam engines. The basic contradiction of the steam-powered vehicle was the relationship of power to weight. The more power, the more weight. The engine should be heavy (to produce enough energy) and the engine should be light (to be manageable on the road without rails). The internal combustion engine resolved rather well the contradiction of the steam engine. The electric car did not. That's why the car with the internal combustion engine won, although the electric car has many other benefits. Now, more than a century later, we are seeing the development of new kinds of electric cars that may finally replace internal combustion automobiles.

Sixth, idle, easily available, but previously ignored resources are used. Resource mapping was studied in Chapter 5. We also can find this requirement in the ideality equation. Benefits can be increased at the same time as cost is decreased only if some new, cheap reserves can be found.

These six evaluation criteria are generic, based on the fundamental concepts of TRIZ. Other criteria are specific to the particular system that is being studied, such as safety, speed of implementation, compatibility with existing systems, compliance with regulatory requirements (which may be different in different countries) or other issues. It is convenient to reserve a place for these miscellaneous criteria. Let's add the seventh criterion: other requirements.

Here are the seven criteria in a list:

1. All harmful features vanish.
2. All useful features are retained and new benefits appear.
3. New harmful features do not appear.
4. The system does not get more complex.
5. The primary tradeoffs and contradictions are removed.
6. Idle, easily available, but previously ignored resources are used.
7. Other requirements related to the developed system are fulfilled.

7.2 MEASURES OF EVALUATION

Cost is not included explicitly in the list of criteria. We have found that, if the idea is a real breakthrough, people will find ways to eliminate cost as a barrier. TRIZ is used repeatedly — first solve the initial problem, then solve the problems of reducing cost. In one recent TRIZ class, people found a way to make a new product for cooking and selling individual portions of food. However, their management rejected the idea because the new factory required would be too expensive, based on the estimate of how much the product would cost and how much they could sell. The TRIZ class was not discouraged. They started a new project that reduced the cost of the proposed factory by 60%. This was enough to persuade their management that the new product could be a success.

Reading this, you may ask how this team was able to reduce the cost of the proposed factory by 60%. This was not the result of one big breakthrough, but rather the result of repeated applications of TRIZ to each of the processes in the proposed factory, focusing on improving the efficiency of each process. In this case, as in many others, we cannot publish fresh examples of good results achieved through the use of TRIZ. A good solution is, by definition, a solution that gives the company a competitive edge. That's why companies such as Procter & Gamble or Ford, both of which have many years' experience with TRIZ, have published very few examples.

In using the list, you will find that comparison of pairs is very clear and much more reliable than attempting to define some absolute level of the concept. New and old technology should also be compared in the same time interval and environment. The ideal lawnmower today is different from the best method of controlling a lawn 5 or 10 years from now, which might be "smart grass" that keeps itself at the right height, rather than a grass-cutting machine. The speed of development of new concepts in the near future is often underestimated. Similarly, ideas for decreasing noise and pollution will undoubtedly get much more valuable in the long run.

Discussion is meaningless if the circumstances for the use of the technology are not defined. There is much discussion, for example, of energy technologies in general. Which is better, solar energy, nuclear energy, coal energy, hydropower or something else? An answer is not possible before establishing for what purpose the energy is needed. Is it for an industrial plant consuming hundreds of megawatts round the clock or for a cottage using some tens of kilowatts temporarily or for a space exploration vehicle enroute to Saturn?

Using the criteria of TRIZ decreases the subjectivity of the evaluation, although it doesn't remove it totally. We don't claim that the set of criteria is 100% comprehensive or that it will mechanically produce an unambiguous result. We stress that it is most fruitful for the development team to discuss the evaluation criteria first and then make the evaluation.

7.3 EXAMPLES OF EVALUATION

The evaluation of new solutions can be easily presented in a table having two columns:

1. Presentation of the criteria, independent of the result (left column)
2. Evaluation (right column)

Let's evaluate some solutions presented in earlier chapters. As we have noted earlier, in these training examples the measures for comparison are well known current technologies. Examples are simplified to make the tool for evaluation as easy to use as possible. Using grass for noise suppression is compared with a big conventional muffler (Table 7.1). The seeding tape is compared with a precision seeder, a simple mechanism with small wheels (Table 7.2). In the case of the latching mechanism, the conical pin is compared with the cylindrical one (Table 7.3). Training embedded in work is compared with traditional classroom training (Table 7.4). Fighting fire with water mist is benchmarked against the use of drops in typical sprinklers (Table 7.5).

Table 7.1 Using Grass and Conventional Muffler to Suppress Lawnmower Noise

Criteria	Comparison with known solution
1. Do the harmful features disappear?	YES Noise will be decreased.
2. Are the useful features retained? Will new benefits appear?	YES It cuts grass as well as the conventional machine.
3. Will new harmful features appear?	NO
4. Does the system become more complex?	NO It gets simpler.
5. Is the inherent primary contradiction resolved?	YES Conflict of big muffler–small muffler is solved.
6. Are idle, easily available, earlier ignored resources used?	YES Grass and geometry used.
7. Other criteria: Easy to implement	YES

Table 7.2   Carrot Cultivation: Seeding Tape Compared with Precision Seeder

Criteria	Comparison with known solution
1. Do the harmful features disappear?	YES System gets simpler; no new devices are needed.
2. Are the useful features retained? Will new benefits appear?	YES Accuracy is retained; speed is improved.
3. Will new harmful features appear?	NO No new harmful features.
4. Does the system become more complex?	NO It gets simpler. No need to buy or rent a machine.
5. Is the inherent, primary contradiction resolved?	YES Conflict of one seed–many seeds is solved.
6. Are idle, easily available, earlier ignored resources used?	YES Strips seed themselves and utilize easily available materials.
7. Other criteria: Easy to implement, usable in all size gardens?	YES

We don't try to claim that the examples present the best possible technologies. Actually, because the examples are proven and published solutions, they are inevitably at least somewhat out of date. However, we are sure this is not a problem. It is great if you have better solutions in mind. Simply insertyour knowledge into the tables for comparison and tailor the examples for yourself. This book provides you with the best possible tools for the generation and evaluation of solutions. You will produce and select the best solutions yourself.

For your own examples, first describe the problem, TRIZ solution and the best conventional solution:

- My problem
- My TRIZ solution
- Best conventional solutions

Table 7.3   Improving Latching Mechanism: Conical Pin Compared with Cylindrical Pin

Criteria	Comparison with known solution
1. Do the harmful features disappear?	YES Yes, there will be no wear.
2. Are the useful features retained? Will new benefits appear?	YES Simplicity retained; the locking problem is solved.
3. Will new harmful features appear?	NO
4. Does the system become more complex?	NO It gets simpler.
5. Is the inherent, primary contradiction resolved?	YES Conflict of big clearance–no clearance is solved.
6. Are idle, easily available, earlier ignored resources used? Is the solution new?	YES The geometry of the pin is used.
7. Other criteria: Easy to implement?	YES

Then fill in the evaluation table (Table 7.6).

7.4 IMPROVEMENT OF THE SOLUTION

If you have the solution near the ideal final result, you should get five "YES" answers (1,2,5,6,7) and two "NO" answers (3,4). But your first idea — even the best basic idea — almost always contains drawbacks that should be removed to get a working solution. After the first formulation of the ideal final result we nearly always find new contradictions and start again on the process of mapping resources to resolve them. The basic concepts of contradiction, resources and ideality form a loop that is repeated many times to develop the new solution (see Figure 7.1).

Using grass as a muffler reduces lawnmower noise considerably. Certainly this solution should be developed further, toward a totally noiseless lawnmower, a lawnmower without toxic exhaust gases and eventually to the nonexistent lawnmower, the garden system that always keeps the grass the proper height.

Table 7.4   Training Incorporated into Work Compared with Traditional Classroom Training

Criteria	Comparison with known solution
1. Do the harmful features disappear?	YES Time is decreased.
2. Are the useful features retained? Will new benefits appear?	YES Training results retained and even improved.
3. Will new harmful features appear?	NO
4. Does the system become more complex?	NO (and a little YES). Good planning makes learning simple for trainees. Some complex work by the training designers is required to set up the system and measure its effectiveness.
5. Is the inherent, primary contradiction resolved?	YES Contradiction lots of time–no time resolved.
6. Are idle, easily available, earlier ignored resources used?	YES Working time is used.
7. Other criteria:	

How about a tape for seeding? There may be seeds in the tape that will not grow. It is necessary to fix seeds on the tape. The tape is a new component. These are new problems whose solutions are necessary to go forward from a plaything stage.

Water-jet cutting, developed in the 1960s, was slowly introduced in industry. Materials could be cut without any wear of tools or generation of excess heat. But these benefits were coupled with an annoying drawback. Slowly moving water is soft. The water must move at high speed to make it act as hard as an abrasive cutter. But high speed requires high pressure. To cut thick, tough materials, pressures of 2000 to 3000 bars are needed. So we have a tradeoff between improved cutting properties and substantially increased complexity of equipment. But suppose we use the water to move microscopic cutting particles such as grains of sand? Then the particles do the cutting and much less water pressure is needed. Indeed, about 20 years after the concept was

Table 7.5 Firefighting: Water Mist Compared with Droplets

Criteria	Comparison with known solution
1. Do the harmful features disappear?	YES Water consumption and damage decreased.
2. Are the useful features retained? Will new benefits appear?	YES Capacity to extinguish fire retained.
3. Will new harmful features appear?	YES No new harmful features.
4. Does the system become more complex?	NO AND YES Machinery for getting high pressure needed.
5. Is the inherent, primary contradiction resolved?	YES Much water–no water solved nearly ideally.
6. Are idle, easily available, earlier ignored resources used?	YES Water is used.
7. Other criteria: Easy to train firefighters to use?	YES

Figure 7.1 The improvement cycle. Repeating the process speeds the development project as a whole and improves the result.

Table 7.6   Evaluation Table

Criteria	Comparison with known solution
1. Do the harmful features disappear?	
2. Are the useful features retained? Will new benefits appear?	
3. Will new harmful features appear?	
4. Does the system become more complex?	
5. Is the inherent, primary contradiction resolved?	
6. Are idle, easily available, earlier ignored resources used?	
7. Other criteria:	

first introduced, abrasive additives were brought in. Jet cutting is now in wide use.

There are many situations in which a rigid system is improved by adding hinges. A lamp has been made more controllable through the use of hinges. The penalty has been more parts. The solution has been further improved by replacing the hinges with elastic components, which can be viewed as many very small hinges. Repeating the cycle of application of the concepts of contradiction, resources and improved ideality made big improvements in the ease of using the lamp. The evaluation criteria play a crucial role in the use of the improvement cycle — your table of YES and NO answers will help you decide what aspect of the problem should be treated as the principal issue on the next cycle.

7.4 SUMMARY

To evaluate any solution, first establish the evaluation criteria and then make the evaluation. The criteria should be, as much as possible, independent of the subjective feelings and interpretations of people. Different people should get approximately the same conclusions using the criteria. The basic concepts of TRIZ are translated into seven evaluation criteria that are used to evaluate proposed solutions with respect to the best conventional solutions.

When a problem is first solved, the evaluation may reveal drawbacks. Don't hurry to reject the idea. It may be bad, but it may also be excellent. There only are sub-problems to solve. The first idea should nearly always be improved. There is a strong psychological barrier preventing the improvement of solutions. Repeating the cycle of application of the concepts of contradiction, resources and improved ideality make big improvements in the solution and help overcome psychological inertia.

8

ENRICHING THE MODEL FOR PROBLEM SOLVING

INTRODUCTION

This book can be divided to three parts: problem solving by analyzing contradictions, the development of systems without direct use of contradiction analysis and the implementation of TRIZ for achieving business objectives. Chapters 2–7 covered the first part:

- Contradiction
- Resources
- The ideal final result

In this short chapter, we reach the midpoint.
Chapters 9–10 contain the second set of concepts and tools:

- Patterns of evolution
- Forty principles of innovation

Chapters 13–15 address the implementation of TRIZ in organizations.

- How to implement TRIZ, the use of TRIZ with various other tools
- Using TRIZ together with the Theory of Constraints (TOC)
- Using TRIZ together with Six Sigma

Figure 8.1 shows a review of the three basic concepts studied so far. It is a more detailed presentation of Figure 7.1.

After the review, we present a general agenda for a problem-solving session. This is a one-page summary of the instructions given in previous

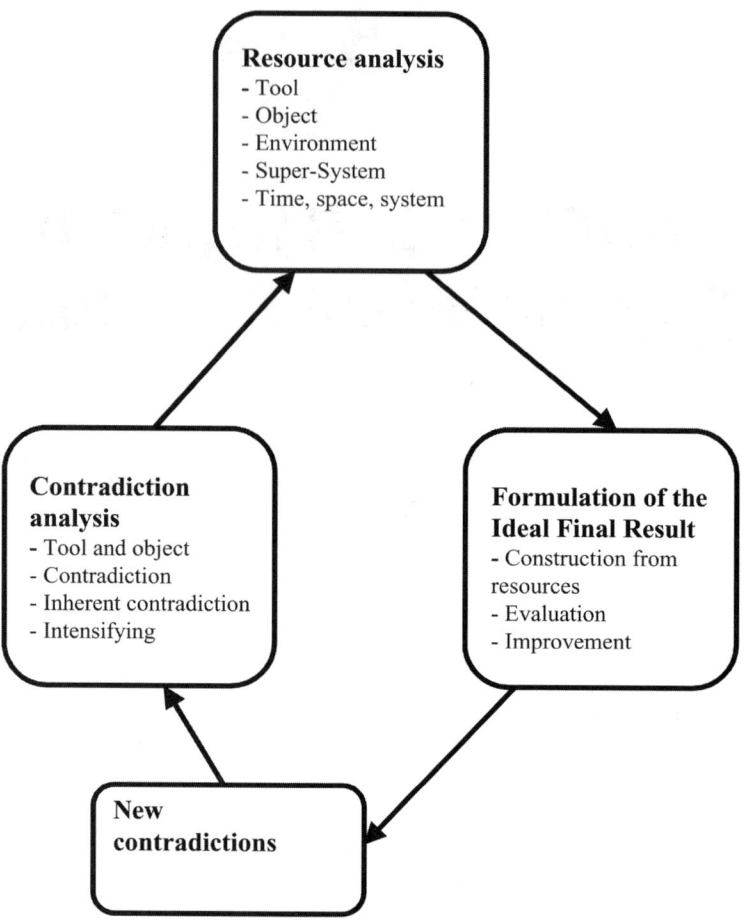

Figure 8.1   The cycle of TRIZ. The three upper boxes present three basic con-cepts. The lowest box shows the reformulation of contradictions. The loop shows that the problem-solving process is repeated. In practical situations, it is not unusual to go around the loop four or more times. You are making progress if you are answering different questions each time.

chapters, which is illustrated with two familiar cases: lawnmower-noise muffling and carrot cultivation (Tables 8.1 and 8.2).

This agenda is a practical method that is easy to teach, easy to learn and will get you started actually using the TRIZ methods fast. Many organizations that are looking for better methods for creativity have resisted TRIZ because of a perception that TRIZ is complicated and difficult. This impression has been due to the long, detailed step-by-step

Table 8.1   Summary of the Lawnmower Example

1   Describe Contradictions (Chapters 3–4)

1.1 Describe the contradictions that make up the problem. There may be several on different system levels and in different stages of the life cycle of the product or process or system.
The lawnmower is too noisy. Contradictions: If noise is decreased, the lawnmower gets more complex. The lawnmower could be eliminated if we had grass that needs no cutting (such as moss), but to develop new grass may take much time and money.

1.2 Select one contradiction to resolve.
If noise is decreased by making the muffler bigger, the lawnmower gets more complex.

1.3 Intensify the contradiction.
Intensified inherent contradiction: *The muffler should be present — the same muffler should be absent.*

2   Map Resources (Chapter 5)

2.1 List resources of the tool and object.
Exhaust gas, noise. Try to use harmful elements as useful resources.

2.2 List resources of the environment.
Air, gravity, the person who pushes the lawnmower.

2.3 List resources on the higher system level (macrolevel) and microlevel
Exhaust tube, lawnmower, grass, soil.

3   Define the Ideal Final Result (Chapters 6–7)

3.1 Remove the contradiction using resources.
Exhaust tube, exhaust gas and grass do the job of the absent muffler. There is no muffler, but the noise vanishes.
One technical solution: grass as muffler.

3.2 Evaluate the solution.
Conflict of big muffler–small muffler is solved.

3.3 Improve the solution.
Imagine, for example, that the exhaust tube shape is changed (maybe the end segmented). The hot gas dries the grass more effectively and gently.

Table 8.2 Summary of the Carrot Cultivation Example

1 Describe Contradictions (Chapters 3–4)

1.1 Describe the contradictions that make up the problem. There may be several on different system levels and in different stages of the life cycle of the product or process or system.

The initial problem: Thinning carrots is an arduous job. This problem could disappear if we could plant seeds very accurately. But then we will have a new problem in another stage of the process: seeding precisely is difficult and time consuming.

1.2 Select one contradiction to resolve.

The more precisely carrot seeds are planted, the slower the speed.

1.3 Intensify the contradiction.

Intensified inherent contradiction: Very many seeds–one seed

2 Map Resources (Chapter 5)

2.1 List resources of the tool and object.

Hand (guided by the eye and brain of the gardener). Seed.

2.2 List resources of the environment.

Soil, water, air, furrow, gravity.

2.3 List resources on the higher system level (macrolevel) and microlevel

Waste paper, waste grass (from lawn).

3 Define the Ideal Final Result (Chapters 6–7)

3.1 Remove the contradiction using resources.

Seeds, soil and water make many seeds into one seed.
One technical solution: biodegradable seed tape.

3.2 Evaluate the solution.

Conflict one-many seeds is solved.

3.3 Improve the solution.

Make the planting process even simpler by adding fertilizer to the seed tape.

Table 8.3   Short Agenda for Problem Solving

1   Describe Contradictions (Chapters 3-4)

    1.1   Describe the contradictions that make up the problem. There may be several on different system levels and in different stages of the life cycle of the product or process or system.

    1.2   Select one contradiction to resolve.

    1.3   Intensify the contradiction.

2   Map Resources (Chapter 5)

    2.1   List resources of the tool and object.

    2.2   List resources of the environment.

    2.3   List resources on the higher system level (macrolevel) and microlevel)

3   Define the Ideal Final Result (Chapters 6–7)

    3.1   Remove the contradiction using resources.

    3.2   Evaluate the solution.

    3.3   Improve the solution.

guides for problem solving and problem statement development in traditional TRIZ teaching systems. The most detailed step-by-step guide for problem solving is ARIZ or Algorithm for Innovative Problem Solving. Altshuller and his team developed different versions of ARIZ between 1956 and 1985. A good review is published in Savransky's book.[1]

In this book, we present a short guide. We believe it is better to know a few things well than many things superficially. A short problem-solving process can be easily repeated and repetition or reiteration is important for mastering a new set of skills.

Use the list in Table 8.3 as the agenda for a problem-solving meeting or as a guide for using TRIZ without meetings. We recommend using this kind of summary of your own examples. Table 8.3 is a template. To show how to use a template we repeat the examples of the lawnmower and carrot cultivation (Tables 8.1 and 8.2).

REFERENCE

1. Savransky, S.D., *Engineering of Creativity*, CRC, Boca Raton, 2000, 304.

9

PATTERNS ARE POWERFUL TOOLS FOR SYSTEM DEVELOPMENT

INTRODUCTION

In Chapter 2, we introduced the patterns of evolution. The purpose of this chapter is to show how to use the patterns. In problem solving, knowing the patterns helps to go from the features of the ideal final result to concrete solutions. In situations where the contradictions are hard to see, understanding the patterns helps you see how the system is evolving. If we see how the system will evolve, we actually know the solution to the problem; in this way, the solution can be developed without contradiction analysis. The place of patterns in the model for problem solving is shown in Figure 9.1.

The five most useful patterns of evolution are the following:

1. Uneven evolution of the parts and the features of the system
2. Transition to the macrolevel or incorporation to the larger system of higher level
3. Transition to the microlevel or the segmentation of the system into smaller parts
4. Increasing the interactions between systems
5. Expansion and convolution of systems

First, each pattern will be explained, and then we'll show how to use them both separately and together.

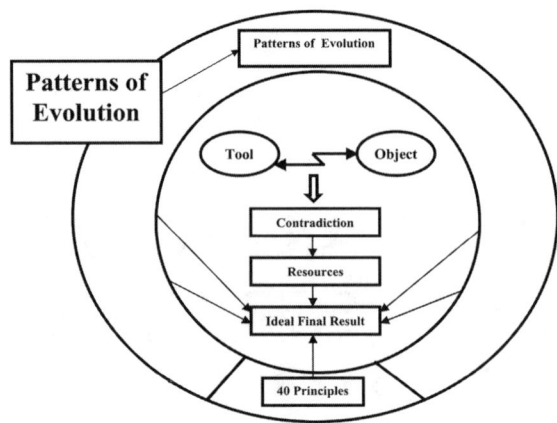

Figure 9.1   Patterns of evolution in the model for problem solving. Patterns are an independent toolkit helping to improve the system — arrows lead directly from patterns to the ideal final result. They also support the problem-solving process from the contradiction to the description of the features of the ideal system.

9.1 THE UNEVEN EVOLUTION OF SYSTEMS

The uneven evolution of the system causes problems, bottlenecks and contradictions all the time. The unevenness concerns all systems and technologies: machinery, processes, organizations, etc. Unlike our typical way of thinking, particularly about technology, the evolution is not linear. Usually current trends are extrapolated directly to future trends. In reality, there are always discontinuities, that is, incremental quantitative changes are broken by qualitative leaps to new technology.

The history of the bicycle is a good example. In 1791 in France, de Sivrac developed the *celerifere*, which had two in-line wheels connected by a beam. The user "drove" it by pushing against the ground with his feet. The "hobby horse" technology was improved throughout the following decades, but there was a bottleneck in the system. The greater the speed, the more difficult it was to move the cycle by kicking the ground. This contradiction was solved when the Michaux brothers added cranks and pedals and created a bicycle boom with their *velocipede*.

Very soon a new contradiction appeared. The greater the speed, the more difficult it was to ride, because the rider had to move his legs faster and faster. Increasing the diameter of the front wheel was the only way to get more speed from one leg motion.

The chain transmission, introduced in 1885, solved the problem of getting higher speed with smaller wheels. Then a new problem developed: the greater the speed, the more vibration. Air-filled tires, which had been invented in 1845, resolved this problem. The bicycle reached the form it has today.

A small exercise: Think about a modern bicycle. Which contradictions you can name? How could they be resolved?

We see that usually some parts, or some features, improve rapidly while other parts and features remain unchanged, sometimes for a very long time. Unevenness appears again and again and compels the system to evolve. The development of the car compelled the building of roads. Better roads made it necessary to develop better cars. (We could argue about whether it was "necessary" to improve the cars, or if it was now possible to sell better cars because there were roads for them and, if there is a market for something new, people will create a product for the market.) Analogously, computer hardware helps to make more effective software useful, and better software compels improvement in the hardware, as anyone knows who has just purchased a new system only to find that it is obsolete 6 months later. See examples and exercise in Figures 9.2 and 9.3.

Exercise: Describe an example of uneven evolution of systems. It could be from your personal life or from your business life.

9.2 TRANSITION TO MACROLEVEL

The pattern of transition to the macrolevel describes a system that becomes better and better integrated into the higher-level system or macrosystem. The system is not developing in vacuum, as an isolated thing, but as part of a larger system.

The bicycle reached some important limits of development near the end of the 19th century. It was not possible to make significant increases in speed and transportation capacity of the human-powered vehicle. The bicycle was integrated together with the internal combustion engine into the higher-level systems. Motorcycles, cars and airplanes developed. The motorcycle is directly a "motorized bicycle." But the car also had its origin not only in horse-drawn carriages, but in bicycles, too. And, after all, the Wright brothers were bicycle mechanics.

Exercise: What other ways can you suggest to integrate the bicycle into higher-level systems?

Stoves or fireplaces for heating a single room were developed to a high level a long time ago. To increase comfort and save time, stoves became central heating systems. In many northern countries, the integration has gone further. Large parts of cities are heated from a single thermal

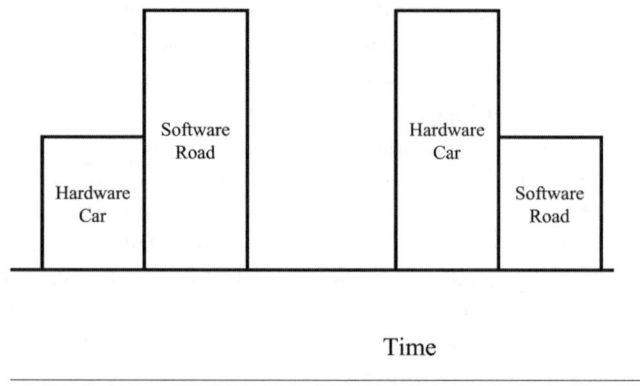

Figure 9.2   The uneven evolution of the systems.

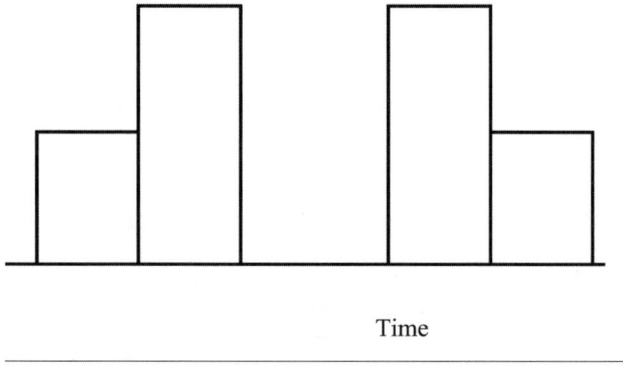

Figure 9.3   Exercise. Illustrate with your own examples the uneven evolution of the system.

power station (so-called district heating, often with co-generation of electric and thermal energy). Analogously, vacuum cleaners are combined into central vacuum cleaning systems.

Clocks have been integrated into radios, TV sets, cars, computers, mobile telephones, microwave ovens and innumerable other systems. Electronic components are integrated, as well as buildings, clothes (many layers) and many food products (for example, the multilayer cake, the casserole with meat, vegetables, starch and sauce).

Many examples of integration can be found in business, marketing, training and other nontechnical fields. The entire financial field of mergers and acquisitions is a mechanism by which companies and other organizations are integrated frequently. Marketing is typically a system consisting of different media and ways of work (marketing mix).

The transition to the macrolevel is such a ubiquitous law that it is often dismissed as trivial. However, many problems that could be avoided arise because this law has been neglected. For example, in the 1970s, Apple's computers (Apple II, Lisa and Macintosh) and Sony's "beta" video systems met difficulties, although they were technically superior as isolated products. But they were inferior due the lack of integration into the larger system that the customer wanted to work with.

Exercise: Describe an example of transition to macrolevel. It could be from your personal life or from your business life.

See also examples and exercises in Figures 9.4 and 9.5.

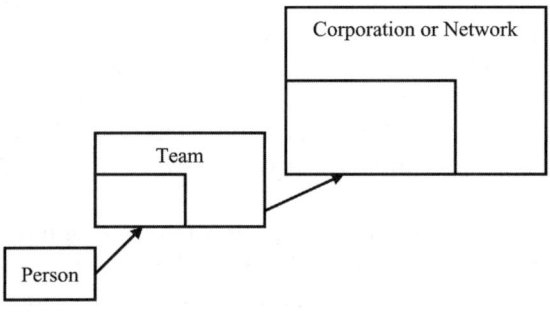

Figure 9.4   Transition to macrolevel.

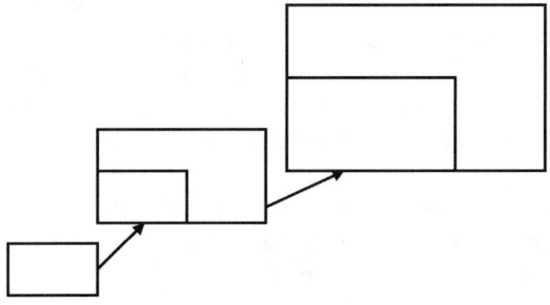

Figure 9.5   Exercise. Illustrate the transition to macrolevel with your own examples.

9.3 TRANSITION TO MICROLEVEL

The pattern of transition to the microlevel describes systems that are improved by dividing them into smaller and smaller parts. In Figure 9.6, three examples of the pattern are presented.

The first example has already been discussed briefly: replacing the solid edge of a cutting tool with a water jet. *Molecules of water, instead of one solid object, do the cutting.* A market research company, Frost and Sullivan, claims that the water-jet tool market has emerged as "the fastest-growing market segment" for the period 1997–2004 (see: http://wj.net/waterjet).

Another example is from medicine. Early stretchers used to transport injured persons were covered with simple canvas or some kind of mattress, which were not very good for carrying injured people with broken necks or backs, because the person could not be held in a rigid position to prevent further injury. Vacuum mattresses were introduced to solve the problem. An air-tight mattress filled with small plastic balls takes the shape of the body of the victim. When the injured person lies on the mattress properly, air is suctioned from the mattress. A vacuum fixes the position of the balls with respect to the patient and each other, holding the patient securely during transportation. *Many little balls replace the single solid support.*

A single big robot can only do a job such as digging or cleaning in one direction. It can be replaced by many mini-robots communicating with each other with radio or infrared waves. *Many tiny robots replace the large one and can do the job from several directions simultaneously.*

Some additional examples:

Washing machines for cars frequently use brushes, but the brushes can scratch the finish of the cars. Now water jets often replace brushes.

Cleaning cloths can go to the microlevel, too. So-called microfiber cloths may be so effective that washing chemicals are not needed.

One process in the working of denim is stone washing to produce the popular look of faded denim. The method is improved by using enzymes instead of stones. This is transition to the microlevel in two ways:

1. The enzyme molecules are much smaller than stones.
2. The enzymes work on the fabric at the molecular level. The stones acted on the level of the threads.

The evolution of printing has gone through many generations of transition to the microlevel. Lithographic printing uses large (200 kg or more) stones. Guttenberg's breakthrough was movable type, with each letter on a separate piece of metal. The matrix printer requires only a few

Examples:

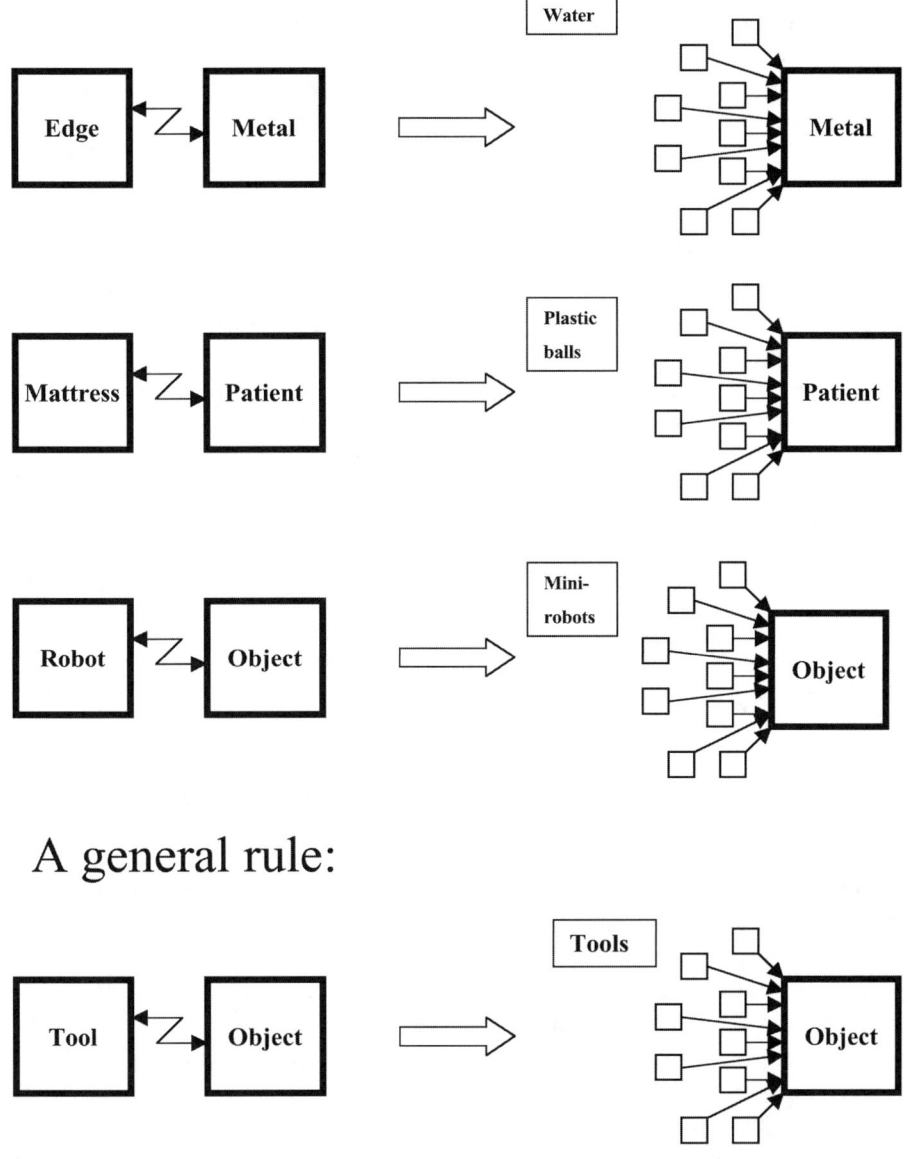

A general rule:

Figure 9.6 Three examples of the evolution pattern "transition to microlevel." Simple tool–action–object models can illustrate many patterns. On the left side is the system with a problem, on the right side the improved system.

(initially nine, later 24) tiny rods to make up each letter. The ink jet printer uses liquid ink and forms the letters from patterns of ink dots (initially 100 dots per inch, now as many as 600 dots per inch) and laser jets use light to sensitize the paper and fine powder to form the letters.

Remember the firefighting example used in several chapters — a fine mist of water replaces liquid water.

A classic example of the transition to microlevel is the manufacturing of glass on melted tin (Pilkington process). Big rolls of hot metal that were used to form the glass plate were replaced by a liquid tin bath used to float the glass plate.

There are three ways to segment material objects:

1. Segmentation of objects: solid body, segmented body, liquid or powder, gas or plasma, field. Most of the examples we have used so far are in this category.
2. Segmentation of space: solid body, hollow body, many caverns, porous substance, pores filled with an active substance. All kinds of spaces inside a body are frequently called "voids."
3. Segmentation of surface: flat surface, corrugated surface, rough surface.

Let's consider some more examples. We have repeatedly used the example of atomized water. If water can be segmented, why not air, too? In many processes, such as purification or flotation, air is mixed with water. Often, processes can be improved using smaller air bubbles, or by using foam.

Transition to the microlevel can be used to some extent in business problems. Huge organizations are often segmented into many small independent organizations to get faster response to customer problems and faster new-product development. The "empowered employee" is a single person acting with the authority of the whole company — to fix problems and to take initiative much faster than the full infrastructure of the company would allow.

In the chapter on resources we used a serialized novel and TV series as examples of segmentation in entertainment.

The best solutions to problems often contain both the transition to macro- and microlevel. In microelectronics and communication technology, segmentation of components has enabled the building of global networks. The evolution of organizations has many analogous features. Small companies and subsidiaries build global networks.

Exercise: Describe an example of transition to the microlevel. It could be from your personal life or from your business life.

See also an exercise in Figure 9.7.

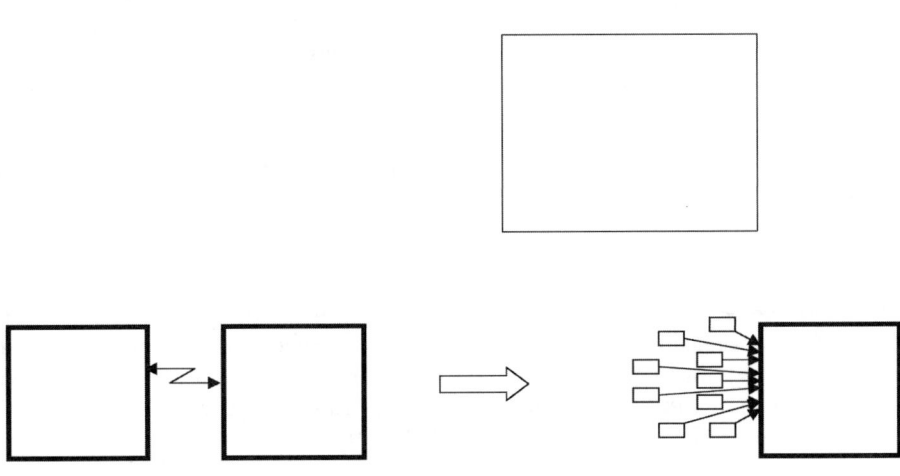

Figure 9.7   Exercise. Illustrate with your own examples the transition to microlevel.

9.4 THE INCREASE OF INTERACTIONS

Increasing interactions means adding new interactions or a transition to better controlled interactions. This pattern also includes adding new substances that interact with the substances in the original system. "Substances" are materials, components, systems and elements. They may be micro-organisms (e.g., yeast), animals (e.g., bees), or humans (e.g., a hand as a tool). Substances interact with each other by a variety of means, including mechanical actions, thermal actions (heat, cold), acoustic interactions (different sounds), chemical reactions, electromagnetic fields and waves, odors and biological interactions. One can find analogies in business: human interactions and communication.

The pattern of increasing interactions can be described in general as follows: The interaction between the tool and the object is insufficient or harmful. The system can be improved by adding new substances to the existing components, adding new interactions, or changing the substances and interactions in a variety of ways to amplify what is insufficient or to eliminate what is harmful.

Consider some examples of the transition to better-controlled interactions (see Figure 9.8).

The interaction between the automobile and its environment is a big problem. New solutions are intensively developed that usually mean the introduction of better-controlled interactions. Navigation systems use radio waves, radar-equipped bumpers, drive-assistance systems with on-board

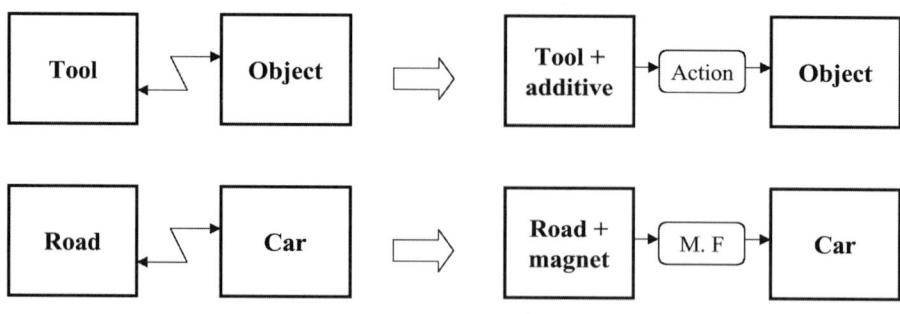

Figure 9.8 Increasing interactions. Mechanical connections are complemented or replaced by more controllable interactions. The mechanical contact between a car and a road can be complemented by electromagnetic interaction (M.F. = magnetic field). Magnetic material is added to the road.

video cameras, etc. In 1997 in California, a magnetic control system was demonstrated. To improve automobile steering, magnetic pins were precisely located in the street. Sensors in the car detected the pins and the steering was modified by the output of the sensor to keep the car from swerving off the road.

The history of the clock is a good example of the transition to more controllable interactions. The first clocks were sundials, which used the shadow of sunlight to indicate time. They could not be used on cloudy days or at night. Sand and water clocks and, later, pendulum clocks, used gravitation. They worked day and night, but were big and awkward. Spring clocks were introduced. They were smaller and easier to use. Modern clocks use vibrations of quartz crystals. The user cannot see the actual time-measuring mechanism. Both the transition to the microlevel and the transition to more controllable interactions are demonstrated in the history of clocks.

More examples:

- Post-it notes fixed by glue (adhesion) replace thumbtacks or pins (mechanical interaction).
- Barbed wire has a long history of improvement for enclosures for cattle. Now, low-energy electrified wire is used in many areas (electric interaction). A recent application of electric wire is to keep bears out of apiaries.
- Ultrasound, at frequencies that humans can't hear, has replaced fences as a way to keep birds out of gardens. This is also an example of the pattern of segmentation, with a field replacing an object.

These examples may help you when considering ways to improve your system by improving interactions.

9.4.1 Simple Introduction of New Substances

The system can be improved by adding a new substance. To improve the performance of steel, carbon or nitrogen is added in the surface layer. To decrease the friction between the hull of an icebreaker and ice, a layer of polymer is added on the hull. Note that this pattern may violate the concept of the use of resources, since it may require the addition of substances that are not resources of the original system. The details of the specific situation will dictate whether new substances are needed, or resources can be used.

9.4.2 Introduction of Modified Substances

Instead of a new substance, one can use a modification of the substances already existing in the system. To improve the performance of steel, the surface layer is quenched. To decrease the friction between the hull of an icebreaker and ice, water is added. The use of modified substances is closer to ideality than the use of new substances, because modifications of the existing resources are used.

9.4.3 Introduction of a Void

Instead of a substance, one can use a void. It sounds odd to say, "Instead of something, use nothing," but that is exactly what we mean. Examples:

- Hollow structures instead of solid ones
- Foamed metal instead of solid metal objects
- Vacuum instead of antibacterial chemicals — a vacuum package
- Use of vacuum and suction in fixing, moving and lifting

A "void" is everything more rarified than its environment.

9.4.4 Introduction of Action

Instead of substances and voids (things and nothing), one can use action. An example: dust can be removed in a cyclone-style vacuum cleaner using a mechanical action — centrifugal force. To improve performance, an electric field can be added to the cyclone.

In the older literature on TRIZ, one can also find the term "field." The study of objects and interactions is called substance-field analysis. In this book, we use the simple model tool–action–object, or tool–interaction–object. The term "interaction" covers both fields in common language (electromagnetic fields, gravitation) and interactions (chemical, thermal, mechanical and biological) that are usually not called fields. One can speak also of social and human interactions. The concept of interaction is accurate and helps the TRIZ user to see opportunities to use different interactions.

Exercise: Describe an example of the increase of interactions. It could be from your personal life or from your business life.

See also an exercise in Figure 9.9.

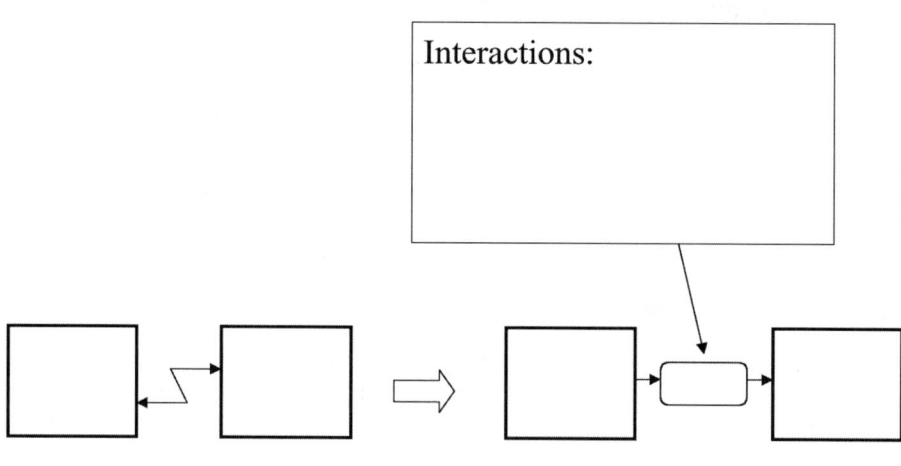

Figure 9.9   Exercise. Illustrate increasing interactions by your own examples.

9.5 EXPANSION AND CONVOLUTION

The last pattern we will consider is called "expansion and convolution," or "trimming," or sometimes "pulsating" evolution. The system expands first, becoming more complicated, then it is trimmed or convolutes; that is, its elements are combined into a simpler system. The increasing number of parts and operations cause problems that are solved when the system is simplified. The evolution is not linear. We can say that the system "pulsates." First, there are few parts and operations. Then the number of components and operation stages grow quickly, until the system "collapses" and is trimmed to a few parts. Then the cycle begins again.

Transistors and other microelectronic components were first used as single parts. Systems with many components became very complex, with short lifetimes. Later, a great number of small components were combined into the integrated circuit, which can be treated as a single component. The electronics industry has been through many cycles, combining integrated circuits into complex systems, then simplifying the system by making components with higher levels of integration. Integrated circuits became a new monosystem that is further embedded into other systems.

But, the pulsating pattern doesn't require high-tech electronics. The traditional bicycle wheel consists of a great number of spokes. Recently, wheels where spokes are combined into a disk or a few spokes have been introduced.

The microfiber cloth used as an example of the transition to the microlevel is also an example of expansion and trimming. The system "cloth-plus-washing-chemical" is trimmed to a microfiber cloth.

The car tire was first improved by adding an inner tube. Later, the inner tube was removed. The spare tire was a staple in automobiles for many years. Now, it is removed in some cars and the tire itself works as the "spare." In heavy trucks, double tires were introduced. There are concept designs with a single very wide tire.

Expanding and trimming also improve processes. Water-jet cutting often makes it possible to merge cutting and machining because the surface doesn't need any machining after water cutting.

Printing technology is evolving from five processes to two: Standard methods require preparing the content, making the film, making the plate or cylinder, fixing the plate/cylinder to the printing machine and printing. The new technology consists of two phases: preparation of the content and printing.

See, too, the citation from Toynbee in Chapter 6: transportation technology expanded first — locomotive — and then is trimmed — internal-combustion engine. Long-distance communication technology first expanded — telegraph and telephone wires — and then is trimmed — wireless technology. Dress expanded first from primitive rags to complex dress in the 17th century and then is simplified (tee shirts and casual slacks.)

This pattern can also be called "mono-bi-poly," because a monosystem is combined with another to form a bisystem, then more are added to form a polysystem. When the polysystem is simplified, it becomes a new monosystem. Salamatov has analyzed the pattern mono-bi-poly in detail.[1]

Both similar and dissimilar systems can be combined. We have already introduced many examples of the mono-bi-poly transition. Fragile glass plates can be handled more easily if they are packaged together. Juice packages can be moved more easily if they are affixed to each other. The word "sandwich" meant at first only slices of bread with a filling. Soon people began calling anything that had multiple layers a "sandwich." Sandwich structures are used in buildings and airplanes. Cloth and textiles, pans and kettles often have many layers. Glass is often sandwiched with other materials — that is more common than using multiple glass layers. One way to make safety glass is to sandwich plastic between layers of glass.

The pattern of mono-bi-poly could be called the sandwich pattern. The combination of similar systems is a special case of sandwich principle.

> **Exercise: Describe an example of the expansion and convolution. It could be from your personal life or from your business life.**

See also examples and exercise in Figures 9.10 and 9.11.

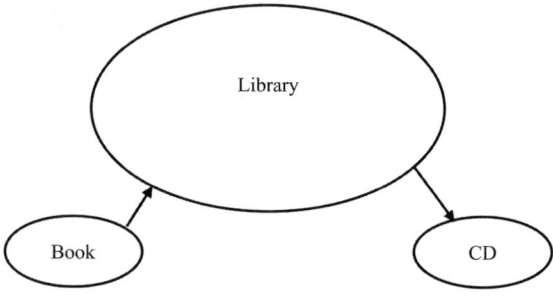

Figure 9.10   Expansion and convolution.

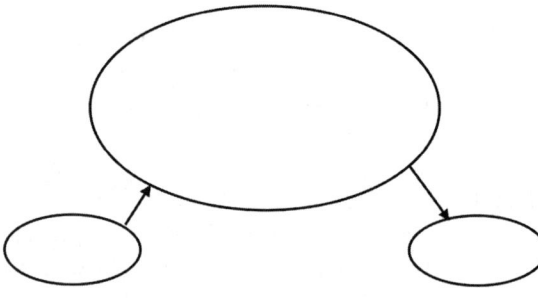

Figure 9.11   Exercise. Illustrate expansion and convolution by your own examples.

9.6 HOW TO USE PATTERNS TOGETHER

Patterns should be studied together and the result checked by the criterion of the ideal final result. Considering only one single pattern can often lead to incorrect ideas about possible patterns of evolution. Many together give much more reliable results. For practical work, we suggest using "five + one patterns," that is: the above-mentioned five patterns plus the pattern of increasing ideality. Table 9.1 summarizes "five + one patterns" and some information about the contents of the patterns.

9.7 BENEFITS FROM UNDERSTANDING THE PATTERNS OF EVOLUTION

We can name at least four uses and benefits of the patterns:

1. Both management and experts can use the patterns of evolution as tools for the evaluation and selection of ideas and solutions to problems. They complement the evaluation by the criteria of ideality. Some good questions to use in reviewing a proposed solution to a problem:
   - Does this solution demonstrate the uneven evolution of the technology? What part will need improvement next?
   - Will the system next transition to the macro- or microlevel?
   - How will interactions increase?
   - How does the ideality of the system as a whole increase?
2. Evolution patterns aid in identifying problems. Examining each pattern can give you information. What are the spearheads and bottlenecks in the evolution? How can you integrate the system into the next higher-level system and how to segment it into smaller parts? How can you increase interactions? One can create "what-if" studies of the future evolution. What can be achieved, if the innate potential of technology is used? When will it be necessary to use a different technology to improve the ideality of the system?
3. The same evolution patterns that help to state problems help to solve them.
4. The use of the solutions from other industries gets easier. It will be easier to see similar features and use them. For example, the segmentation and integration principles that are used in electronics and machine building can be transferred to the building industry. Patterns of self-service in retail business are repeated in education and in medicine.

Table 9.1   A Summary of Patterns

Pattern	About the pattern
Uneven evolution of the system	Uneven evolution of parts. Uneven evolution of process stages. Uneven improvement of features. Repeating rise of unevenness.
Transition to macrolevel	One system is combined with a similar or a dissimilar system, or with many similar or dissimilar systems (mono-bi-poly). Transition to macrolevel is repeated.
Transition to microlevel	Solid body, segmented body, liquid or powder, gas or plasma, field. Solid body, hollow body, many caverns, porous substance. Flat surface, corrugated surface, rough surface.
The increase of interactions: introducing substances and actions	Introducing substances: new and modified substances, void. Introducing actions: mechanical, acoustic, thermal, chemical, electric, magnetic.
Expansion and convolution	Increasing number of parts. Increasing number of operations. Convolution to fewer parts and operations. Cycles of expansion and convolution are repeated.
Increasing of the ideality of the system	One pattern is used to increase ideality. If the use of one pattern causes new problems, other patterns are used to resolve them. Many patterns are used.

9.8 EXAMPLES OF THE APPLICATION OF EVOLUTION PATTERNS

New solutions can be obtained and known systems improved by applying the patterns of evolution.

Tables 9.2 and 9.3 summarize the application of the patterns of evolution to two cases used as examples earlier in this book. Table 9.4 is the exercise of application.

Table 9.2   Decreasing the Noise in the Lawnmower

Pattern	How to apply to the system
Uneven evolution of the system	Noise suppression, decreasing pollution.
Transition to macrolevel	Combination with other machines. Solving the problem at the level of the garden system.
Transition to microlevel	Porous materials.
The increase of interactions	Noise against noise. Sensing the sound waves and generating waves that cancel them.
Expansion and trimming	Increasing and decreasing number of parts. First making the muffler bigger, later eliminating the muffler altogether.
Summary: increasing of ideality	Cleaner lawnmower. Less lawnmower. Absent lawnmower. Grass stays short by itself.

9.9 SOME NUANCES IN THE USE OF PATTERNS

We have presented the patterns in simple ways and omitted many details. But to avoid oversimplification, keep these points in mind:

■ Uneven evolution of the system, the transition to macrolevel, and pulsating evolution are the most universal patterns. By "universal" we mean that, in almost every case, we can see these patterns and use them to develop the system.

Table 9.3 How to Cultivate Carrots

Pattern	How to apply to the system
Uneven evolution of the system	Thinning and seeding, with much manual work, are backward operations in gardening.
Transition to macrolevel	Adding fertilizers to the tape.
Transition to microlevel	Imagine many microcarrots instead of a few big ones.
The increase of interactions	Collecting sunlight.
Expansion and trimming	Adding the tape.
Summary: increasing of ideality	Single action needed to plant at the right spacing.

Table 9.4 Select One of Your Problems and Apply the Patterns To It

Pattern	How to apply to the system
Uneven evolution of the system	
Transition to macrolevel	
Transition to microlevel	
The increase of interactions	
Expansion and trimming	
Summary: increasing of ideality	

■ The pattern of increasing interactions is perhaps the most "statistical" of those considered. For example, systems contain more and more electric and magnetic interactions. The trend has been steady for more than 100 years. However, this does not mean that mechanical interactions should always be replaced or compete with electric and magnetic fields. The pattern says that the transition to better controllable fields happens so often that this possibility should be considered.

■ The transition to the microlevel happens frequently, too. Some few exceptions are found. For example, sometimes chemical washing is replaced by mechanical cleaning, or cleaning by water or steam to get rid of chemicals. One should remember the probabilistic character of patterns and check whether the changes implied by them increases the ideality of the system.

One frequently asked question is how one can speak of any patterns or "laws" in the evolution of systems, when predictions of the future and society are very unreliable. If the patterns are true scientific laws, shouldn't we be capable of precisely predicting the evolution, at least the evolution of technology? The question contains a misinterpretation. The most rigorous and exact sciences, such as physics, tell what will happen if all the necessary conditions are met. But, science doesn't predict when society will choose to meet all those conditions. Physics enables us to calculate how much time and energy are needed to travel to Mars. But it cannot tell precisely when humans will land on Mars. The patterns of evolution tell *what* can be achieved, *if* resources are devoted to the task.

There is also the time factor. The longer the time period considered, the more regularity one can see and of less important is the influence of occasional factors and subjective decisions. In the beginning of the book, we presented examples of "late" innovations. Obviously innovations like penicillin, fast-food restaurants and flash melting of metals could appear some time earlier, or some time later than they really happened. In the long run, the evolution was inevitable. One can accelerate or retard the change, but not prevent it.

9.10 SUMMARY

■ The same patterns are repeated in the evolution of systems. These patterns can be used for the further development of the system.
■ Five primary patterns of evolution: uneven evolution of the technology, transition to macrolevel, transition to microlevel or segmentation, increase of interactions and expansion/convolution or trimming.
■ Use the patterns of evolution for selecting solutions, finding or solving problems, forecasting evolution, transferring solutions across industries.

REFERENCE

1. Salamatov, Y., *TRIZ: The Right Solution at the Right Time*, Insytec, Hattem, 1999, 192.

10

PRINCIPLES FOR INNOVATION: 40 WAYS TO CREATE GOOD SOLUTIONS

INTRODUCTION

We have repeated throughout this book that understanding the common features of good solutions is crucial to the enhancement of creativity. One obvious conclusion is to simply make a list of the most important features and then use the list for generating bright ideas and successful products.

Good generic solutions across industries have been studied as part of the development of TRIZ. Altshuller and his researchers collected examples of repeated use of the same solutions from patent information. After painstaking work, the information on tens of thousands of good solutions was boiled down to 40 principles in the early 1970s.[1] The use of these principles of innovation became an important branch of the theory. Various collections of standard solutions and principles for innovations were developed. To learn more about the research, see Savransky's book[2] and the paper by Zlotin and Zusman.[3]

Our goal in this chapter is to present the modern list of 40 principles as a problem-solving tool that is effective, easy to use, cheap and accessible to everybody. The 40 principles are the most popular tool of TRIZ and several books and many articles are devoted to them. This version has three main features:

1. All examples are new. They are mainly examples of innovations or realized solutions that are actually used in everyday life.
2. Most principles are illustrated with examples from both business and technology.
3. The principles are presented in a compact form, without division to subprinciples, as in early books. One can manage 40 principles much easier than 80–90 subprinciples.

Some of the features of earlier versions of the 40 principles are preserved in this version:

1. The structure of the list of the principles, that is, the number (40 — not less, not more) and the order of principles are the same as in older books, so this book is compatible with older publications.
2. The names or labels of the principles are also conventional. For example, one principle is named "strong oxidants." We show how, for some situations, this can mean the use of strong emotions, as well as the use of particular chemicals, but we keep the classical name for the principle.

Research continues and there may possibly be more than 40 principles at some time in the future (or fewer, if the list is reorganized.) For now, "40 principles" is the standard and new examples and new ways of using them have been added to expand their use. If you would like to get lots of additional examples to help you solve your problems, try any of the following issues of *The TRIZ Journal*:

■ Business Examples: September 1999
■ Social Examples: June 2001
■ Architecture Examples: July 2001
■ Food Technology Examples: October 2001
■ Software Examples: September and November 2001

The structure of this chapter is very simple. First, we will introduce the principles with some examples. Then we will show how to select principles that will be most likely to help solve particular problems, including the contradiction matrix as an important tool for the search.

Figure 10.1 shows the place of the principles in the model for problem solving. You can use the 40 principles as a totally independent tool. This chapter is written so that you can read it and get benefits from it without reading the other chapters of the book.

The efficiency of the 40 principles is increased when they are used together with other tools. Patterns of evolution, ideality and contradiction

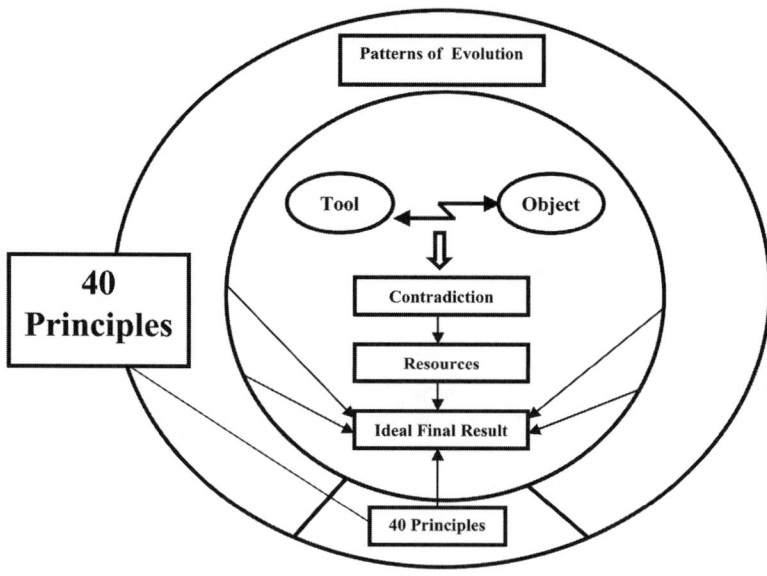

Figure 10.1 Forty principles in the model for problem solving. The arrow from the box "40 Principles" to "Ideal Final Result" means that you can use principles as an independent tool. You can also use it for the development of the result after the analysis of contradictions and resources. Both shortcuts and longer ways are available in the model.

analysis may give the same solutions or may give different solutions from the 40 principles — considerable overlap is quite common. The tools strengthen and enrich each other.

10.1 THE GENERAL REVIEW OF 40 PRINCIPLES

Below is a list of all 40 principles. Alternate names come from a variety of translations of the original Russian research.[4]

1. Segmentation (fragmentation)
2. Separation (taking out, extracting)
3. Local quality
4. Symmetry change (asymmetry)
5. Merging (consolidation)
6. Multifunctionality (universality)
7. "Nested doll" (Nesting, "Matrushka")
8. Weight compensation (anti-weight, counterweight)
9. Preliminary counteraction (preliminary anti-action, prior counter-action)
10. Preliminary action (prior action, do it in advance)

11. Beforehand compensation (beforehand cushioning, cushion in advance)
12. Equipotentiality (bring things to the same level)
13. "The other way around" (do it in reverse, do it inversely)
14. Curvature increase (spheroidality, spheroidality-curvature)
15. Dynamic parts (dynamicity, dynamization, dynamics)
16. Partial or excessive actions (do a little less)
17. Dimensionality change (another dimension)
18. Mechanical vibration
19. Periodic action
20. Continuity of useful action
21. Hurrying (skipping, rushing through)
22. "Blessing in disguise" (convert harm into benefit)
23. Feedback
24. Intermediary (mediator)
25. Self-service
26. Copying
27. Cheap disposables
28. Mechanical interaction substitution (use of fields)
29. Pneumatics and hydraulics
30. Flexible shells and thin films
31. Porous materials
32. Optical property changes (changing the color)
33. Homogeneity
34. Discarding and recovering
35. Parameter changes (transformation of properties)
36. Phase transitions
37. Thermal expansion
38. Strong oxidants (accelerated oxidation)
39. Inert atmosphere (inert environment)
40. Composite materials

The use of each of the principles is illustrated by examples from several different areas of technology and business. Many examples that were used earlier in the book are repeated here, to show how the 40 principles can be used to develop solutions to those problems. Problems can be solved and systems improved different ways, using one principle or using several together. In most solutions, more than one principle is used. When you find an interesting principle, look for other principles that can improve the idea. Often, one principle will give you a concept for a solution, but several may be necessary to get to a practical working solution.

To make it easier to read and remember, the list of principles is divided into groups of two to four. Each group is considered in one section. The principles in some groups are naturally connected with each other; others are simply lists of different approaches. The groups are:

- Segmentation, separation (Principles 1–2)
- Local quality, symmetry change, merging, multifunctionality (3–6)
- "Nested doll," weight compensation (7–8)
- Preliminary counteraction, preliminary action, beforehand compensation (9–11)
- Equipotentiality, "the other way around," curvature increase (12–14)
- Dynamic parts, partial or excessive actions, dimensionality change, mechanical vibration (15–18)
- Periodic action, continuity of useful action, hurrying (19–21)
- "Blessing in disguise," feedback, intermediary (22–24)
- Self-service, copying, cheap disposables, mechanic interaction substitution (25–28)
- Pneumatics and hydraulics, flexible shells and thin films, porous materials (29–31)
- Optical property changes, homogeneity, discarding and recovering (32–34)
- Parameter changes, phase transitions, thermal expansion (35–37)
- Strong oxidants, inert atmosphere, composite materials (38–40)

10.2 SEGMENTATION, SEPARATION (1–2)

10.2.1 Principle 1

Segmentation. Fragmentation. Transition to microlevel. Divide an object or system into independent parts. Make an object easy to disassemble. Increase the degree of fragmentation or segmentation (see Figure 10.2).

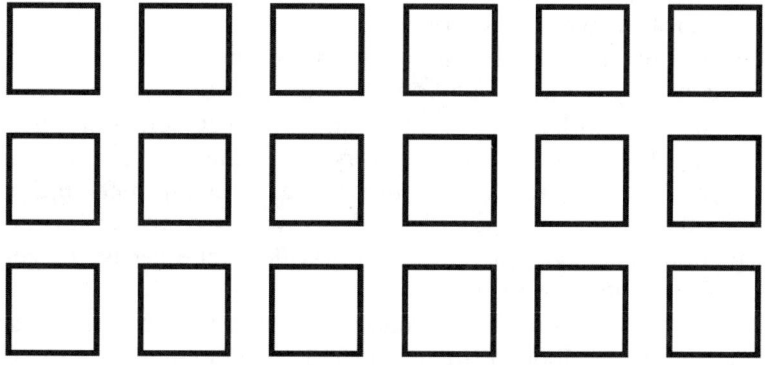

Figure 10.2   Principle 1. Segmentation. Divide a system into parts.

Examples:

- The law of the transition to microlevel, considered in the previous chapter, is the result of the repeated application of the principle of segmentation. Segmentation is a very frequently used principle. It helps to "combine the incompatible" and meet contradictory requirements in many different problems. Recall the example of fire extinguishing. More effective firefighting (good) requires more water and causes water damage (bad). Water is necessary and, at the same time, there should be no water. Let's segment water to small droplets and then increase the degree of fragmentation to mist. Mist can suppress fire effectively, using a very small amount of water. In previous chapters, we processed the problem through many steps. It is possible to get the idea directly, using the single principle of segmentation.

- Stone washing is one of the technologies used to give denim needed features and appearance. Stones are, however, rather crude tools and clumsy machinery is needed to handle them. A nice solution is to use enzymes instead of stones to get the same result. Enzymes are not small stones, but the concept of using molecules instead of large objects comes from the principles of segmentation.

- In previous chapters, we considered the noise problem of the lawnmower. It is useful to imagine different ways to segment or fragment the lawnmower, the muffler and its environment. In the chapter on resources (Chapter 5), we speculated about using many small automatic mini-mowers instead of one big lawnmower. Sun-activated automatic lawnmowers, wandering over the lawn like sheep, have already been developed. In other applications, mini-robots are in use (for example, for examination and cleaning of tubes). Perhaps some day the idea of the automatic lawnmower and mini-robots will be combined.

- While waiting for the noiseless lawnmower, let's try to segment the muffler. The idea of using grass as the muffler (see earlier chapters) can be seen as an example of segmentation also. Obviously, this is not the only way of segmentation. Why not try replacing a single exhaust tube by many small ducts and a single muffler with many small ones? New materials and production technologies, such as casting of plastic, make it easy to get many components in one.

- If we can segment the lawnmower, why not increase the segmentation of grass? Indeed, there are mulching lawnmowers that cut the grass into very small pieces. The benefit is that small pieces of grass will degrade and fertilize the soil and there is no need to remove them from the lawn. This solves a different problem — waste removal instead of noise reduction — but it is common in TRIZ to find new opportunities for improvement.

■ Carrot cultivation is another example considered earlier. How can we use the segmentation principle here? We can imagine many mini-carrots instead one big plant. They can grow very densely, almost without soil, if they get water and necessary fertilizers (compare with hydroponics or cultivation in water). There are already gardening technologies for growing a greater number of little plants and getting a larger overall crop from the same area.

■ In Chapter 5 (resources) we also discussed the possible segmentation of the pin in the latching mechanism. The shape of a layered or filament pin can be controlled better than the shape of a solid component.

■ The segmentation principle has many applications in business. The segmentation of the market is a common practice and a commonly used term. Most big corporations have segmented themselves into business units or profit centers. A corporation should be small to be flexible and big to have enough resources for production and marketing. To be small and big at the same time, a huge company is divided into subsidiaries, profit centers or other units working relatively independently. The ABB Company and the Gore Corporation are exemplary — they create a new organization whenever an existing part of the company exceeds 150 people.

■ For the past 30 years, the use of teams has been one of the persistent themes in the workplace, because small teams are flexible and can make decisions quickly.

■ A large job can be broken into many smaller jobs (called a work breakdown structure in project management). The JIT (just-in-time or Kanban) system uses the concept of segmentation to an extreme — it replaces the idea of mass production with the idea that the most efficient production system can produce a single unit just as easily as multiple units.

■ Entertainment examples, such as serialized novels and TV movies, were considered previously. A new entertainment medium is emerging on the Internet, where a novel is not only serialized, but has several options in each chapter, so the reader can select the segments for a personalized book. This combines segmentation with Principle 3, local quality.

■ Segmentation is a good and often accessible way to use resources. The system can be divided into smaller parts. New components and substances are not needed.

More examples across industries:

■ Make cupcakes instead of one large cake, so people can decorate according to their own tastes and a variety of flavors can be offered.
■ Use powdered welding metal instead of foil or rod to get better penetration of a joint.
■ Inject a drug in a finely powdered form.
■ Use Java applets. The predecessor of this example was the use of piping in Unix to make large tasks possible by combining sequences of small tasks.
■ Paper is traditionally coated by a transferring application by a blade on a web. A new way is to spray coating in atomized form at high speed on both sides of the paper web.

10.2.2 Principle 2

Separation. Separate the only necessary part (or property) or remove an interfering part or property from an object or system (See Figure 10.3).

We usually need only a part of the system or some property or feature. Examples:

■ We need light, not lighting devices. Today, many parts of the lighting system are located some distance away from the places lighted. Reflectors and fiber optics both can separate the mechanism, such as a lamp, from the point of use of the light. Fiber optics are used in tiny surgical instruments to provide light exactly where it is needed inside the surgical area without the bulk of a lamp.
■ We don't need a vacuum cleaner as such, but cleaning capacity. A central vacuum cleaning system leaves only nozzles and a piece of tubing in the apartment. Noisy and dirty parts are located where they don't disturb inhabitants.
■ An electric lawnmower can work quite well if the lawn is not too large. The production of energy is removed from the lawn.

Figure 10.3   Principle 2. Separation. Separate a part or property from a system.

- Put a noisy air-conditioning compressor outside a building and pipe the compressed air to the place where it is needed — most medical and dental offices are built this way.
- The example of fighting fire with mist illustrates the separation principle. Only small-diameter tubes are needed at the fire's location. The heavy part of the equipment is removed.
- Franchising separates the ownership of a local business such as a restaurant or a printing shop from the development of the concept and the systems that make it successful.

Does the paper producer need the paper-making machinery? Does an insurance company need mainframe computers and data storage? Only a few years ago, this question didn't deserve attention. The insurance company outsources operation of its data center to a specialist company and the paper company outsources operation and maintenance of the mill. The ASP (application system provider) is a new business concept — many companies don't own their own software, but rent it as needed from a provider.

Do we need personal cars? Bicycles? Until recently, the answer was "yes" for convenience. Today, new shared-use schemes are emerging, at least on an experimental level. The user buys the right to use a car or a bicycle for a certain time. For cars, the credit card works as the key. In Portland oregon, the bicycle experiment requires the user to put the bicycle where the next person who needs one can take it. These experiments give us transportation capacity, individual routes and schedules, comfort, prestige, certain life-style and many other features, but *separate* the use of the automobile from the ownership, care and cost of a ton of metal and plastic. The proponents of these new schemes claim that a customer can actually have many cars by not owning any. Today I can use a small vehicle in the city, tomorrow a big car for a long trip and the day after tomorrow a limousine for prestige purposes. Costs can be cut, because of the more intensive use of the capital invested in the automobiles.

Often we have a contradiction between present and absent. Some awkward machinery or complex process should be present to get a needed feature and the same machinery or process should be absent to save space, energy and time. The separation principle may be a solution in problems of this sort.

More examples:

- Use a recording of a barking dog, without the dog, as a burglar alarm. Likewise, use the sound of birds in distress instead of a scarecrow.
- Use fiber optics or a light pipe to separate the hot light source from the location where light is needed.

■ Outsource maintenance and operation services.

Exercise: Think of one example from your personal life or your business for each of the principles in this section.

1.Segmentation	
2.Separation	

10.3 LOCAL QUALITY, SYMMETRY CHANGE, MERGING AND MULTIFUNCTIONALITY (3–6)

10.3.1 Principle 3

Local quality. Change an object's structure or an external environment (or external influence) so that the object will have different features or influences in different places or situations. Make each part of an object or system function in conditions most suitable for its operation. Make each part of an object fulfill a different and useful function (see Figure 10.4).

Often the object should have some additional feature, but the introduction of this feature causes new problems or makes the system more complex and expensive. We should change the system and we should not change it.

It is easier to change the system locally. There are many examples in technology. Quenching and other treatments of the surface layer of metal components make the surface properties different from the bulk properties of the material. We use a different wrench for every nut, because fixed-size wrenches are much stronger than adjustable wrenches. Specialized compartments in a lunchbox for each type of food keep hot things hot, cold things cold and make it safe and economical for workers and schoolchildren to carry their lunches with them.

In business, the segmentation of the market also illustrates the local quality principle. *Segmentation* is used to divide the market into small markets with specific attributes, then *local quality* is used to treat each of those markets appropriately. To tailor its approach in the automatic washing machine market to the cultural preferences of each group, the Whirlpool Company has hired marketing people in India who speak 18 different languages.

Figure 10.4 Principle 3. Local quality. Change an object's structure.

Local quality applies to people, as well. Some are most effective working on their own and others are most effective in teams. Intensive professional specialization is needed for certain skills and a broad liberal arts background is required in other situations.

More examples:

- Precision farming using the correct amount of chemicals where needed
- Pencil with eraser
- Hammer with nail puller
- Kids' areas in restaurants

10.3.2 Principle 4

Symmetry change. Change the shape of an object or system from symmetrical to asymmetrical. If an object is asymmetrical, increase its degree of asymmetry (see Figure 10.5). Asymmetric paddles mix more effectively than symmetrical ones (both for concrete and for cake batter). Asymmetric scissors are handier than symmetric ones.

Increasing asymmetry is a way to use geometric resources. It is often easier to change the geometry than to introduce new substances or components.

Mass customization is a business strategy that corresponds to asymmetry — the product or service or policies of a business are specifically designed for each customer and don't need to be the same as those provided to other customers.[5,6]

More examples:

- Use astigmatic optics to merge colors.
- Budget for different departments individually rather than using a constant percentage increase or reduction for all departments.

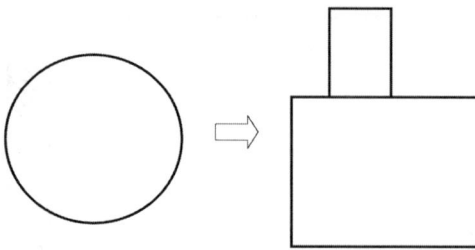

Figure 10.5 Principle 4. Symmetry change. Change the system from symmetrical to asymmetrical.

10.3.3 Principle 5

Merging. Bring closer together (or merge) identical or similar objects; assemble identical or similar parts to perform parallel operations. Make operations contiguous or parallel; bring them together in time (see Figure 10.6).

Examples of merging:

- Integration in micro-electronics.
- Telephone and computer networks.
- Paper sheets constitute a book; books are merged into a library.
- See examples of carrot seeding and of the handling of packages in previous chapters. Seeds can be placed precisely and quickly if fixed in position on the tape. Small packages can be moved easily if they are affixed to each other.
- Fragile and weak components, such as glass plates, can be made stronger without increasing weight by combining them into packages.
- An array of radio telescopes has greater resolving power than a single dish.
- An idea to get rid of traffic jams is to make vehicles travel in small "platoons" under computer control. A motorway lane could handle about 6,000 vehicles in hour instead of today's 2,000.

If an object should be small *and* big and if there should be many *and* one, merging is often a solution.

In the section on segmentation (Principle 1), we discussed large corporations that are dividing themselves to get small at the same time. Small companies or individual entrepreneurs often have the opposite problem: how to get big while remaining small. Networking is perhaps the most popular solution. Others are: chains of companies, franchising schemes and conventional mergers. Segmentation and merging principles

Figure 10.6   Principle 5. Merging. Bring closer together similar objects.

are often most effective if used together. Organizations are segmented and then the parts merge. (But keep the principle of local quality in mind. The use of one single principle may lead to an ineffective or wrong solution.) Two or three principles together may work better than any one alone. In the 1970s, E.F. Schumacher launched the slogan "small is beautiful." Later the slogan was forgotten, because big can be just as beautiful.

The pattern of the transition to the macrolevel (Chapter 9) is the result of repeated applications of the principle of merging and the pattern of expansion and convolution is the result of alternating merging with separation and with next principle, multi-functionality.

10.3.4 Principle 6

Multifunctionality or universality. Make a part of an object or system perform multiple functions; eliminate the need for other parts (see Figure 10.7). The number of parts and operations decreases and useful features and functions are retained.

Examples:

- ABB has developed an electric generator with high voltage. A conventional transformer is not needed because the generator can directly feed the electric network. In some new-car designs, the flywheel, alternator, starter and some other parts are combined into a single component. A single adjustable wrench for all nuts is another example.
- People use the universality principle also. Cross-functional training makes people much less susceptible to layoffs, because they have multiple skills instead of one skill. Compare this with the local quality principle.

More examples:

- Team leader acts as recorder and timekeeper.
- One-stop shopping — supermarkets sell insurance, banking services, fuel, newspapers, etc.
- Handle of a toothbrush contains toothpaste.

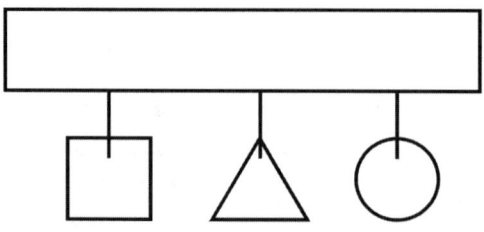

Figure 10.7   Principle 6. Multifunctionality. Make a system perform multiple functions.

Exercise: Think of one example from your personal life or your business for each of the principles in this section.

3. Local quality	
4. Symmetry change	
5. Merging	
6. Multi-functionality	

10.4 NESTED DOLL AND WEIGHT COMPENSATION (7–8)

10.4.1 Principle 7

Nested doll. Place one object inside another; place each object, in turn, inside the other. Make one part pass through a cavity in the other. The name of this principle comes from the Russian folk art dolls or *matrushkas*, in which a series of wooden dolls are nested one inside the other (see Figure 10.8).

Examples:

■ The double hull in oil tankers
■ Telescoping structures (umbrella handles, radio antennas, pointers)

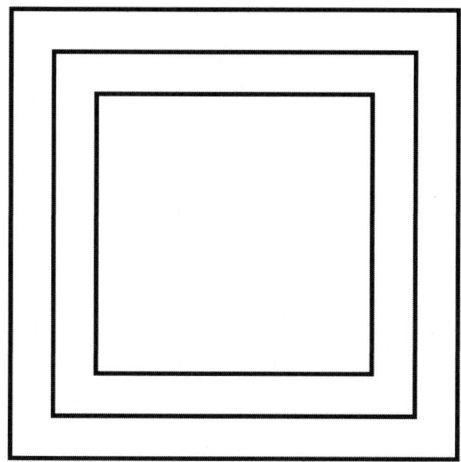

Figure 10.8   Principle 7. "Nested doll." Place one object inside another.

- Business analogies: a special exhibit for one designer inside a boutique store inside a big market
- File structures in the Windows computer operating system (e.g., Chapter 10 is a file in the NewTRIZBook file, which is a file in the My Documents file, etc, …)
- Measuring cups or spoons
- Stuffing a turkey with sausage, stuffing the sausage with chestnuts, etc.

10.4.2 Principle 8

Weight compensation. To compensate for the weight of an object or system, merge it with other objects that provide lift. To compensate for the weight of an object, make it interact with the environment (e.g., use aerodynamic, hydrodynamic, buoyancy and other forces). See Figure 10.9.
   Examples:

- Air tanks in submarine vessels.
- Lifting bodies — the shape of the fuselage acts like a wing and generates lift. Used in both aircraft and ship design.
- Banners and signs cut so that the wind lifts them for display.
- Business analogies: compensation for the heavy organization pyramid with project organization, process organization, temporary organization and other less hierarchical systems "lift" the heavy structure.
- Helium balloon used to support advertising signs.
- Companies increase flagging sales by making connections with other rising products (e.g., movie tie-ins)

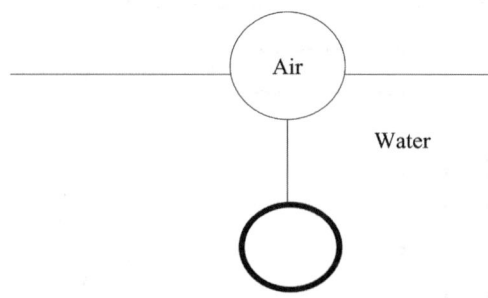

Figure 10.9 Principle 8. Weight compensation. To compensate for the weight of an object or system, merge it with other objects that provide lift.

Exercise: Think of one example from your personal life or your business for each of the principles in this section.

7. Nested doll	
8. Weight compensation	

10.5 PRELIMINARY COUNTERACTION, ACTION AND BEFOREHAND COMPENSATION (9–11)

10.5.1 Principle 9

Preliminary counteraction. If it will be necessary to do an action with both harmful and useful effects, this action should be replaced with anti-actions (counteractions) to control the harmful effects. Create stresses in an object or system that will oppose known undesirable working stresses later on in time (see Figure 10.10).

Examples:

- Use an electric heater to preheat the car engine before starting in the winter in northern regions. Damage to the engine from running with frozen oil is prevented, fuel is saved and air pollution decreased.
- Pre-tense rebar before pouring concrete for stronger structures.

- Changes and innovations usually meet resistance in an organization. Get the affected people involved so they can participate in the planning of changes and don't feel threatened.
- Use customer trials to launch high risk new products (e.g., film companies shoot several endings to a movie and try them with different audiences before final selection).
- Focus on pro-action instead of reaction in maintenance.

10.5.2 Principle 10

Preliminary action. Perform, before it is needed, the required change of an object or system (either fully or partially). Pre-arrange objects so that they can come into action from the most convenient place and without losing time for their delivery (see Figure 10.11).

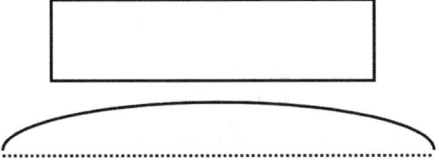

Figure 10.10   Principle 9. Preliminary counteraction. Introduce counteractions to control harmful effects.

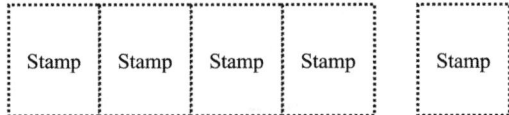

Figure 10.11   Principle 10. Preliminary action. Perform, before it is needed, the required change of a system. Examples: Sheets of stamps are perforated before sale, to make separation easy. Food stores were arranged before the expedition to North and South Poles.

Examples:

- Preliminary perforated packaging is easy to open.
- Precut parts for the building of wooden houses save work at the construction site.
- Workers arrange their workspace so the most frequently used tools (either physical, paper or electronic) are the easiest to reach.

- Do market research, study possible futures, build reserves for changes.
- Prepare solutions to problems customers aren't complaining about today, but may notice tomorrow: environment, moral code.
- TV demonstration chefs always have neat little dishes with all ingredients premeasured.

10.5.3 Principle 11

Beforehand compensation. Prepare emergency means beforehand to compensate for the relatively low reliability of an object or system over time (see Figure 10.12).

Figure 10.12   Principle 11. Beforehand compensation. Prepare emergency means beforehand to compensate for the low reliability of a system.

Well-known technological examples are airbags in cars and pressure relief valves in boilers and chemical reactors.

Non-technical examples are posting instructions for possible emergency situations: fires, the use of narcotics among personnel, environment problems and preparation of equipment (first aid and rescue kits, fire extinguishers) where they may be needed.

The frequently asked questions (FAQs) sections of many Web sites are examples of Principle 11 — commonly, users are told how to help themselves solve problems that are known to exist in the system.

Principles 9–11 together constitute a group of time-related principles, either for preventing problems or for correcting them quickly.

Exercise: Think of one example from your personal life or your business for each of the principles in this section.

9. Preliminary counteraction	
10. Preliminary action	
11. Beforehand compensation	

10.6 EQUIPOTENTIALITY, THE OTHER WAY AROUND AND CURVATURE INCREASE (12–14)

10.6.1 Principle 12

Equipotentiality. Change operating conditions to eliminate the need to work against a potential field (e.g., eliminate the need to raise or lower objects in a gravity field). See Figure 10.13.

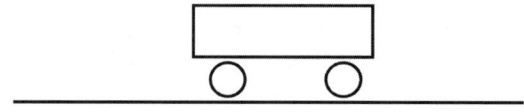

Figure 10.13 Principle 12. Equipotentiality. Eliminate the need to raise or lower objects or "walk uphill" some way.

In the flat factory, high shelving one does not practically raise or lower components.

Use spring systems to lift sheets of wood to the right height so workers can slide them into the machine for the next step in the process. The Bishamon Company makes these devices — when sold as productivity tools, they were a moderate success. When advertised as tools to prevent workers' back injuries, sales increased dramatically.

Use grounding straps to bring people and objects to equal electrical potential, to prevent harm from static electricity.

A business analogy might be a transition to a flatter organization with fewer hierarchical layers. One step in team formation is to bring all team members to the same level — eliminate hierarchical behaviors.

10.6.2 Principle 13

The other way around. Invert the action(s) used to solve the problem (e.g., instead of cooling an object, heat it). Make movable parts (or the external environment) fixed and fixed parts movable. Turn the object (or process) upside down (see Figure 10.14).

Sometimes this principle is applied very literally — turning machines upside down has solved many industrial problems. One core principle of effective design for manufacturing and assembly is to let gravity be your friend — always position parts so they will fall naturally into the desired place. Turn a carton so that the label can be applied on the top or turn an assembly so that screws can be inserted from the top.

Examples:

- Slow food instead of fast food
- Work at home instead of increasing travel time
- Customers find their own answers in a consultant's database instead of having the consultant find the answer for them.
- Television and radio bring church services to people at home, instead of people going to the church.

When considering an innovative principle, think also of the opposite idea. An industrial plant can be improved by making it flat, according to Principle 8. The need to raise and lower objects is eliminated (Principle 12). Sometimes, it may be better to build very many stories, according to Principle 17: dimensionality change. Any number of approaches, of course, can be combined.

Urban planning provides a mixed technological and social example, illustrating inverted approaches. The garden-city movement called for small cities with low buildings. Erecting skyscrapers is another approach. Le

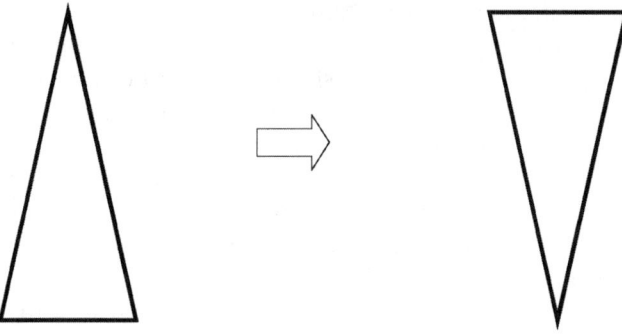

Figure 10.14 Principle 13. "The other way around." Turn the object or process "upside down".

Corbusier's Radiant City combines both schemes. The center has a very high population density and 95% of the land remains free of construction. More examples:

- Benchmark against the worst instead of the best
- Expansion instead of contraction during recession
- Corporate unlearning — acquiring the ability to forget about the past where appropriate
- A small screen in front of each eye (virtual reality monitor) instead of one big screen

10.6.3 Principle 14

Curvature increase. Instead of using square, rectangular, cubical or flat parts, surfaces or forms, use curved or rounded ones; move from flat surfaces to spherical ones; from cube or parallelepiped shapes to ball-shaped structures. Use rollers, balls, spirals or domes. Go from linear to rotary motion. Use centrifugal forces (see Figure 10.15).

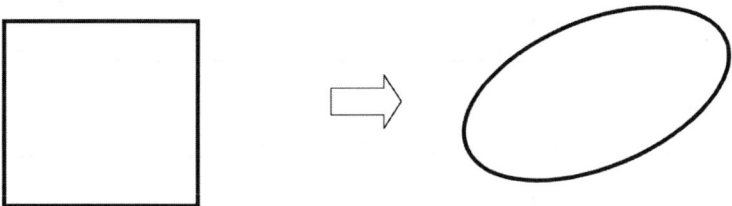

Figure 10.15 Principle 14. Curvature increase. Move from rectangular forms to curved ones.

The filament in incandescent lamps was initially straight. Efficiency improved when it was coiled. Photography was first done with flat plates of glass coated with sensitive emulsion. Cameras became portable when rolls of film were developed. The transition from flat surfaces to curved ones can be easily seen in cars and telephones. A mowing machine for agriculture started with a saw-like reciprocating edge. New machines have rotating blades similar to those in lawnmowers.

Corrugated forms often improve strength without increasing weight. Rotary motion frequently makes equipment simpler.

The principle of curvature increase is often paired with Principle 4: asymmetry. Components can be improved by making them more symmetric or more asymmetric. Curvature increase may increase or decrease symmetry.

Some non-technical analogies can also be found. For example, increasing circulation of information benefits organizational function. Curved walls and streets make neighborhoods visually identifiable (both in cities and inside large office buildings and schools).

More examples:

- Fuller's geodesic dome does away with bulky beams.
- Spherical casters are used instead of cylindrical wheels to move furniture.
- Rotate leadership of the team.
- The Dyson vacuum cleaner spins dirt at high speed, forcing dust outward to the wall. The bag is not needed.
- Use iteration and design loops.

Exercise: Think of one example from your personal life or your business for each of the principles in this section.

:

12. Equipotentiality	
14. Curvature increase	

10.7 DYNAMIC PARTS, PARTIAL OR EXCESSIVE ACTIONS, DIMENSIONALITY CHANGE, MECHANICAL VIBRATION (15–18)

10.7.1 Principle 15

Dynamic parts. Allow (or design) the characteristics of an object, external environment, process or system to change to be optimal or to find an optimal operating condition. Divide an object or system into parts capable of movement relative to each other. If an object (or process or system) is rigid or inflexible, make it movable or adaptive (see Figure 10.16).

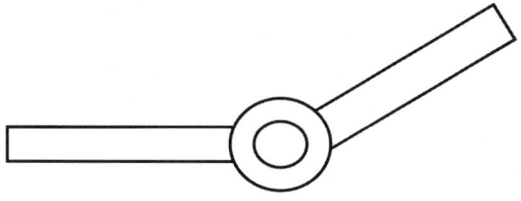

Figure 10.16 Principle 15. Dynamic parts. Make a system or process movable or adaptive.

Repeated use of this principle and combination with the principle of segmentation results in the pattern of increasing interactions and the pattern of transition to a microlevel. Some steps in increasing dynamics:

- Rigid, immobile system
- One hinge
- Many hinges
- Elastic system
- A field instead of a physical object or system

In Chapter 7 we presented an example of a lamp that was made more controllable by introducing hinges. The penalty has been an increasing number of parts. The solution has been further improved by transition to elastic components. The single elastic component has many microlevel parts, very many, very small hinges. Here Principle 1 (segmentation) helps the system to get more dynamic. Generally, if the improvement by one principle causes new difficulties, involve a different principle to solve the new problem.

Stiff and immovable structures are often replaced by more dynamic ones: flexible printed circuits and accumulator batteries in electronics, flexible and self-breaking light poles on the roadsides, wings that change form in airplanes (through the use of flaps and slats on fixed wing aircraft and through motion of the wings on fighter aircraft) and other dynamic structures.

A first step to make a building safe for earthquakes was to make it more rigid: thicker walls, for example. Later, to avoid impractical heavy structures, certain dynamics were added: the building now has bearings and shock absorbers that allow it to move a little. Strength is increased without extra weight. The automobile has gone through similar evolution: first safety belts (passenger fixed more stiffly), then a dynamic part, the airbag, was added.

In business, flexibility — the capability to make changes when the environment changes — is often the difference between success and failure. Organizations are also evolving from rigid, unchanging structures to flexible ones. Ways to increase flexibility are segmentation, flatter organizations (see Principle 12), preparing changes before encountering a problem (Principles 9–11), discarding and recovering (Principle 34) and others.

Schools, too, have used the principle of dynamics as part of their improvement strategy. In many schools, students are no longer assigned to a fixed grade in which all 8-year-olds do third-grade studies together. Rather, the curriculum is flexible. The author's nephew recently was doing fifth-grade arithmetic, third-grade language studies and a personal project to learn geography, all on the same day, in a program that was based on his abilities and interests.

More examples:

- NASA has successfully tested an aircraft that can inflate its wings in flight.
- Software applications have user-configurable toolbars and interfaces.
- "Cafeteria" benefits allow employees to pick which types of insurance, health coverage, etc., they want.
- Traditional printed traffic signs are often replaced by signs that vary in response to changes in weather and traffic conditions.
- In self-righting vessels, the ballast shifts, making the vessel right itself.
- Smart traffic lights help drivers to arrive when the lights are green and discourage frequent starts and stops.
- Flexible manufacturing systems (FMSs) are increasingly used in industry.
- To prevent hydroplaning, speed limits on highways can be adjusted, for example, according to the thickness of the water layer on the pavement.

10.7.3 Principle 16

Partial or excessive actions. If 100% of a goal is hard to achieve using a given solution method, the problem may be considerably easier to solve by using slightly less or slightly more of the same method (see Figure 10.17).

Figure 10.17   Principle 16. Partial or excessive actions. Make slightly less or slightly more.

A classic example is to dip a brush in paint to acquire excess paint, then letting the excess drip off. Similarly, attach a stencil to a surface to be painted, then paint the whole thing. When the stencil is removed, the goal will be achieved and the stencil will take the excess paint with it.

Perforated packages are easy to open. (Cut a little bit, don't cut the whole thing.)

Preparing sketches and concepts helps many writers get finished results more quickly.

If marketing cannot reach all possible customers, a solution may be to select the subgroup with the highest density of prospective buyers and concentrate efforts on them. Another solution is an excessive action: broadcast advertising will reach many people who are not potential buyers, but the target audience will be included in the group that is reached.

10.7.3 Principle 17

Dimensionality change. Move an object or system in two- or three-dimensional space. Use a multistory arrangement of objects instead of a single-story arrangement. Tilt or reorient the object, lay it on its side, use its other side (see Figure 10.18).

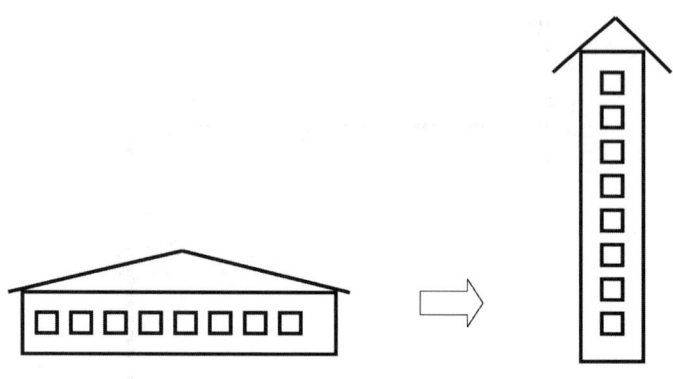

Figure 10.18   Principle 17. Dimensionality change. Move an object or system in two- or three-dimensional space.

In our earlier example of the cultivation of carrots, plants are placed in rows — that is, one dimensionally. Square-foot gardening is a method developed for small gardens. Plants are placed in square blocks (a 4-foot block subdivided into 16 1-foot squares); that is, two dimensionally. They are placed much closer than in large-scale agriculture. The yield is higher and weeds almost nonexistent because of the close spacing.

Use of underground tunnels and buildings is increasing. At the same time, more and more high buildings, multilevel highways and overpasses are built. In winter, Toronto becomes a multistory city — people can travel more than 5 miles indoors by going from the basement of one building to enclosed walkways at the third floor of another, to the main lobby of the next. See also an earlier example of a garbage bin: vertical dimension is used to get space.

Sometimes the additional dimension is invisible to the customer. Disney World in Florida pioneered the multidimensional concept now used in many amusement parks. A network of tunnels, workshops, dressing rooms, storerooms and staff centers runs under the park. Characters are never seen in part of the park that doesn't match their role — instead, the worker vanishes from one area and uses the underground system to travel to the new area or to rest or remove or replace parts of a costume. This preserves the "magic" for visitors. Less glamorously, garbage is dumped into another system of tunnels, so visitors never see garbage being transported through the park.

More examples:

- Holograms as three-dimensional photographs
- IMAX movies with three-dimensional effect

10.7.4 Principle 18

Mechanical vibration. Cause an object or system to oscillate or vibrate. Increase the frequency of vibration. Use an object's resonant frequency. Use piezoelectric instead of mechanical vibrators. Use combined ultrasonic and electromagnetic field oscillation (see Figure 10.19).

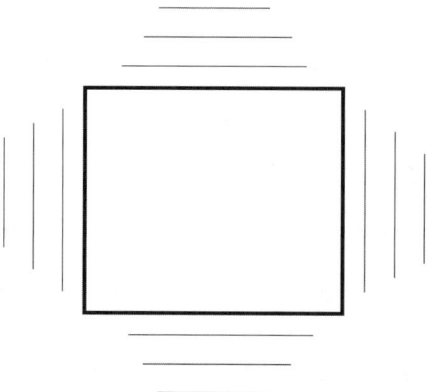

Figure 10.19   Principle 18. Mechanical vibration. Cause a system to vibrate.

Technological examples:

- Vibration instead of sound can be used to alert someone of an incoming call or message on a mobile telephone or pager.
- An object's resonant frequency is used for destruction of stones in the gallbladder or kidneys by ultrasound in a technique called lithotripsy, which makes surgery unnecessary. This can also be seen as the use of segmentation, because the stone breaks itself into very small pieces that the body then eliminates through its natural processes.
- Coordination is an analog of mechanical vibration similarly applicable in organizations. An example: working time and transportation schedules can be staggered and coordinated to decrease traffic congestion. Some people have also used vibration as a metaphor for putting a system into an excited state and then applied Principle

18 to various ways of exciting people to get coordinated action — examples range from cheerleaders or doing "the wave" at sporting events, playing music at political rallies, etc.

Exercise: Think of one example from your personal life or your business for each of the principles in this section.

15. Dynamic parts	
16. Partial or excessive actions	
17. Dimensionality change	
6. Mechanical vibration	

10.8 PERIODIC ACTION, CONTINUITY OF USEFUL ACTION AND HURRYING (19–21)

10.8.1 Principle 19

Figure 10.20 Principle 19. Periodic action. Use periodic or pulsating actions.

Periodic action. Instead of continuous actions, use periodic or pulsating actions. If an action is already periodic, alter the periodic magnitude or frequency. Use pauses between impulses to perform a different action (see Figure 10.20).

Examples:

■ In Whirlpool's washing machine, the pump pulsates. The company claims that the resulting wave effect removes dirt 40–60% more effectively than a conventional agitator.

- Instead of a continuous light signal, a flashing light is often used for information, advertising and warning.
- Researchers propose that taking naps in the middle of day will increase the efficiency of intellectual work.
- Electrical energy for lighting and work is used most heavily in the daytime. Using financial incentives, power companies are attempting to get people to change the amplitude of the periodic action by moving power consumption to nighttime. In some warehouses, electric forklifts are programmed to recharge themselves between 2 a.m. and 5 a.m. because that is when power is least expensive.
- Pauses in work can be used for training.

10.8.2 Principle 20

Continuity of useful action. Carry on work continuously; make all parts of an object or system work at full load all the time. Eliminate all idle or intermittent actions. Note that these last two principles contradict each other — if you eliminate all intermittent actions, you won't have any pauses to use. This just emphasizes that the various suggestions in each principle must be applied with common sense to a particular situation (see Figure 10.21).

The changing character of manufacturing shows considerable influence of this principle. Lean and just-in-time manufacturing methods both emphasize small, customized production runs instead of long series.

Figure 10.21   Principle 20. Continuity of useful action. Carry on work continuously.

More examples:

- Continuous casting of steel and other metals
- Study during traveling
- Continuously variable automatic transmission (CVT)

See also Principle 19, periodic action, and Principle 21, hurrying.

Mechanical typewriters produced all lines in the same direction (depending on the language being written) with no writing during the time it took to return the carriage to the starting position. Electric typewriters with memory astonished the world when they showed the increased productivity of writing in both directions, thus eliminating the pauses between lines.

10.8.3 Principle 21

Hurrying or skipping. Conduct a process or certain stages (e.g., destructive, harmful or hazardous operations) at high speed (see Figure 10.22).
  Examples:

- During surgery, the longer a patient is anesthetized, the higher the risk of failure and future complications. Open-heart surgery that once took 8 hours or more is now done in less than 1 hour, using combinations of new tools and methods. The injection gun is another example from medicine: "Instead of allowing drugs to permeate gradually through the skin, new injection methods force them into the bloodstream at such high speeds that they pass straight through the outer layers of the skin. These pain-free injections do not use needles — they rely on gas pressure to inject the drug, which is prepared in a finely powdered form ... Since the guns do not penetrate the skin, they do not have to be discarded after a single use for reasons of hygiene."[7]

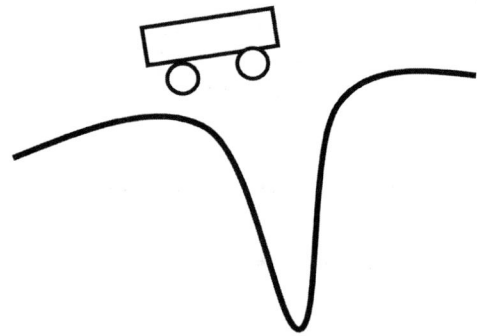

Figure 10.22 Principle 21. Hurrying. Conduct a process at high speed.

- A classic example of this principle is cutting plastic pipe very quickly. If you cut it slowly, heat from the cut region will propagate into the rest of the pipe, making it change shape.
- The traditional method for pasteurizing milk is to heat it to 72°C (161°F) for 15 seconds. Ultra-high-temperature pasteurization, in which milk is heated to 138°C (280°F), for only 2 seconds, increases the storage time of the product.
- In business, it may sometimes be more important to act quickly than to slowly produce perfect work. Thomas J. Watson, IBM

founder, put it as follows: "If you want to succeed, double your failure rate." The importance of being first to market in order to establish a new standard has been emphasized in many studies of the e-business economy.

Periodic action (Principle 19), continuous action (20) and hurrying (21) together compose a useful and practical set of tools. Keep them all in mind. The point is to apply the proper principle to the situation.

Agriculture is an illustrative example from technology. Seeding and fertilizing are typical periodic actions, following which the crop grows continuously. Hurrying is a good word to describe harvesting, because the crop is at its perfect state of ripeness for a very short time.

Project management and personal time management are examples from business. Sometimes hurrying is the most reasonable way (write a report or letter from the beginning to the end without pauses). If the job is big, it can be done only in parts (periodic action).

More examples:

- Fast cycle–full participation method: involve the whole organization simultaneously in a major change such as a company reorganization.
- Get the live lobster into the pot very quickly, holding the lid in your other hand, to prevent being splashed with boiling water.

Exercise: Think of one example from your personal life or your business for each of the principles in this section.

19. Periodic action	
20. Continuity of useful action	
21. Hurrying	

10.9 BLESSING IN DISGUISE, FEEDBACK AND INTERMEDIARY (22–24)

10.9.1 Principle 22

A *blessing in disguise* or "turn lemons into lemonade." Use harmful factors (particularly harmful effects of the environment or surroundings) to achieve a positive effect. Eliminate the primary harmful action by adding it to another harmful action to resolve the problem. Amplify a harmful factor to such a degree that it is no longer harmful (see Figure 10.23).

Examples:

- Electric charges that are usually harmful can be used for the control of process. Charges can cause fires and explosions, destroy electronic components, make materials stick and do other harm, but, the same charges, present in most industrial processes, can give valuable information for the optimization of process. In coal firing, the measurement of electrical charges on coal dust is used to control burning, decrease harmful nitrogen oxide emissions and improve efficiency. In cement and lime grinding, the product quality can be improved.
- Thermal expansion is often harmful and requires compensating devices, but sometimes, it can be used to make a strong and reliable joint without fixing components. See Principle 37 (thermal expansion).
- In an organization, complaints and destructive critique are negative "charges" that can be modified and used to bring about positive change.
- Singing the blues is an example also. The singer turns personal hardship into entertainment for others.
- Virus attacks in computer networks are never good, but each time the system survives an attack, information is generated that makes the system better protected from the next attack. This is similar to the way the body works — surviving an illness generates antibodies that protect the victim from the next incipient infection.

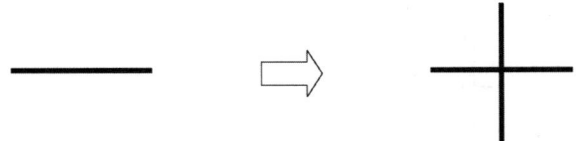

Figure 10.23   Principle 22. "Blessing in disguise." Use harmful factors to achieve a positive effect.

- In many situations, people have turned lemons into lemonade. The founder of TRIZ and author of the 40 principles, Genrich Altshuller, began his work in the 1940s in Baku, in the former USSR, at the time of Stalin. The government "rewarded" his activity by some years in work camps. There, Altshuller met many highly qualified experts who had been arrested during the great purges in the 1930s. He asked them to give classes and seminars. He established a "university" with one student and many professors. Thus, Altshuller obtained encyclopedic knowledge that he used to develop tools for problem solving.

How can we amplify a harmful factor so that it becomes less harmful or becomes useful? Think about the harmful factor as a resource in a different process. Sulfur in coal or other fuel is a harmful component, because the purification of the sulfur dioxide from exhaust gases requires complex technology and the unpurified gas is poisonous to people. However, if the sulfur content is increased, the production of sulfuric acid can become profitable. In the technology of flash-melting of copper (an example in the first chapter), sulfur is used both as fuel and raw material for the production of sulfuric acid.

During the 1940s, one problem for combat aircraft was the potential explosion of gasoline when gasoline fumes mixed with air in partially empty tanks. Carrying an inert gas to fill the empty space in the tanks would require extra weight and complexity. In that case, another resource was used — the exhaust gases from the engine were produced onboard the aircraft and had much less oxygen than ordinary air. The exhaust gases were pumped into the tank and prevented the explosive mixture from forming.

Setting backfires is a well-known technique for fighting forest fires. Controlled fires are set ahead of the fire being attacked to use up the fuel. When the fire reaches that area, it goes out.

More examples:

- Vaccination is a classic example of how to make harmful viruses to protect humans against themselves.
- Build up a tolerance to the allergen by exposure to an extract of the same allergen.
- Unhealthy salt (NaCl) combined with bad-tasting potassium chloride (KCl) creates a healthy, good-tasting table salt.

10.9.2 Principle 23

Feedback. Introduce feedback to improve a process or action. If feedback is already used, change its magnitude or influence (see Figure 10.24).

A technical example: in a typical car, a driver makes observations and uses the steering mechanism, brakes and other "actuators" to make necessary corrections. In new designs under development, the car has active

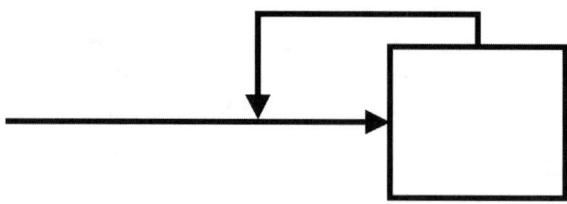

Figure 10.24 Principle 23. Feedback. Introduce feedback.

driver-assistance systems with feedback. If the driver, for example, takes a curve too fast, the system turns the steering wheel automatically.

The evolution of measurements and control is another example. On-line measurements and on-line control are increased. Quality control in production is improved by introducing the immediate measurement and control during the production process, rather than inspection after production. In business, systems for getting feedback from customers are being continuously improved.

Feedback is a primary learning mechanism. Both babies and adults use it naturally, without thinking. People try something new. They examine the result. If it was successful, they do it again. If it was not successful, they modify it, then try it again. This applies to people learning TRIZ from this book and to a baby learning to walk and to all other kinds of learning.

Another example:

■ Heart rate monitors for controlling intensity during exercise

10.9.3 Principle 24

Intermediary. Use an intermediary carrier article or intermediary process. Merge one object temporarily with another that can be easily removed (see Figure 10.25).

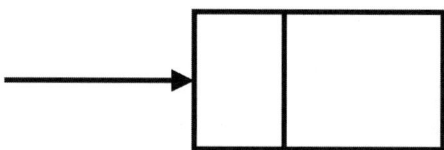

Figure 10.25 Principle 24. Intermediary. Use an intermediary carrier article or intermediary process.

Examples:

- Fixtures or jigs are used to position parts to make assembly easy. The assembled parts are removed and the jigs are used over and over again. This concept is applied in the home as well as in the factory — a baking pan or a gelatin mold is an "intermediary" for shaping a dessert, while a potholder is an intermediary for carrying a hot dish to the table without burning the server's hands.
- Ice can be used to hold small components in place temporarily if they will not be harmed by water when the ice melts. See also the earlier example of fixing carrot seeds on tape.
- A neutral third party can be used as an intermediary during difficult negotiations. For sales promotion, an intermediary who is seen by the customer as an impartial expert can make recommendations.

Exercise: Think of one example from your personal life or your business for each of the principles in this section.

22. Blessing in disguise	
23. Feedback	
24. Intermediary	

10.10 SELF-SERVICE, COPYING, CHEAP DISPOSABLES AND MECHANICAL INTERACTION SUBSTITUTION (25–28)

10.10.1 Principle 25

Self-service. Make an object or system serve itself by performing auxiliary helpful functions. Use resources, including energy and materials — especially those that were originally wasted — to enhance the system (see Figure 10.26).

Examples:

- In a tire that repairs itself, liquids are sprayed inside the tire. When the tire is punctured, the liquid fills the hole. When it contacts the outside air, it solidifies, forming a permanent repair.

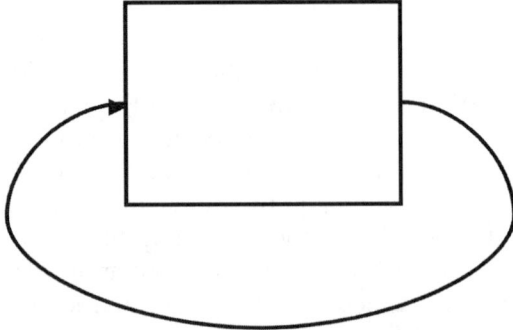

Figure 10.26   Principle 25. Self-service. Make a system serve itself.

- A classical example of self-service in business was presented in the beginning of the book: a self-service fast-food restaurant. Many electronic business ideas are based on including the customer and the customer's resources in the system as resources of the system — this includes everything from communities of interest and chat rooms to data exchanges such as Napster and Gnutella.
- Some search engines use the frequency of use of a Web site as the indicator of quality, so the more often a site is used, the higher it rates on their recommendation list. This is a combination of feedback (24) and self-service.
- Self-treatment and self-test. Patients themselves can perform some medical tests, like the measurement of blood pressure or blood sugar or testing for fertility (then later testing for pregnancy) previously done only by medical personnel. In some cases, patients also adjust their treatment or behavior based on the test results.

Self-service is a way to use the object's resources. This principle illustrates the pattern of increasing ideality. What is more ideal than a system serving itself?

More examples:

- Halogen lamps regenerate the filament during use — evaporated material is redeposited.
- Lend out temporarily underutilized workers to other organizations. (Load-capacity balancing across companies creates a win–win situation where the worker [or player, in the case of football teams]

stays match fit, the lender saves wages and the lendee fills the skill shortage).

- Self-charging quartz watch is powered by the wearer's movement.
- Self-righting lifeboat can capsize and right itself again.
- Modern technology of melting steel scrap uses the energy of scrap itself; carbon and silicon are burned.

10.10.2 Principle 26

Copying. Instead of an unavailable, expensive or fragile object, use simpler, inexpensive copies. Replace an object or system or process with optical copies. If visible optical copies are already used, change the wavelength to infrared or ultraviolet (see Figure 10.27).

Examples:

- Make measurements from an image instead of directly. This includes a wide spectrum of technologies, from satellite photographs of farm and timber resources to ultrasonic images of a fetus in the womb.
- Use a simulation instead of the object. This applies to many business processes as well as to products and services.
- Use prototypes for testing new systems so that any harm is detected early.
- Use virtual prototypes instead of physical ones.
- Use video-conferencing instead of travel.
- Use virtual reality to test new processes or to train people to do work in difficult situations. Surgeons now test new operating procedures on virtual patients and automobile assembly workers practice new procedures in virtual factories.
- Scan rare historic books, documents, etc., so they can be made accessible to all, while the original remains protected.
- Use telepresence instead of fully independent robots.

Fake furs and leathers are also examples of the copying principle. Artificial grass might be an acceptable alternative in some places. No need to mow. Compare with plastic flowers and plants.

Ultraviolet light shows certain kinds of skin lesions better than visible light; dyes that are sensitive to ultraviolet are used to find cracks in metal parts. Infrared images show the heat — this is the basis of most night-vision systems.

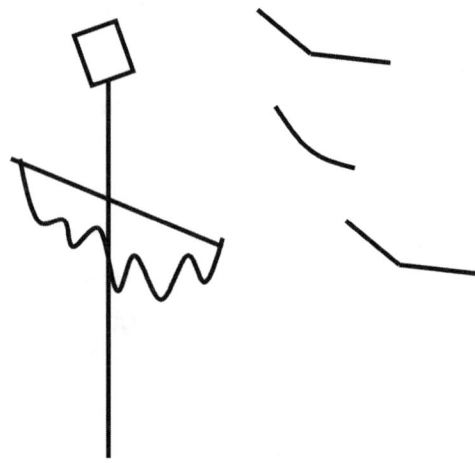

Figure 10.27 Principle 26. Copying. Use inexpensive copies. For example, a scarecrow instead of a person.

10.10.3 Principle 27

Cheap disposables. Replace an expensive object with multiple inexpensive objects, compromising certain qualities (such as service life, for instance. see Figure 10.28).

- Disposable paper and plastic tableware
- Disposable surgical instruments
- Disposable protective clothing

10.10.4 Principle 28

Mechanical interaction substitution. Replace a mechanical method with a sensory (optical, acoustic, taste or smell) method. Use electric, magnetic and electromagnetic fields to interact with the object. Change from static to movable fields to those having structure. Use fields in conjunction with field-activated (e.g., ferromagnetic) particles (see Figure 10.29).
Examples:

- The best-known example of the use of a smell as a warning is the incorporation of bad odors into natural gas to warn users when the system has a leak.
- The just-in-time manufacturing systems use Kanban cards or objects such as portable bins to indicate visibly when supplies are needed.

Figure 10.28   Principle 27. Cheap disposables. Replace an expensive object with multiple inexpensive objects.

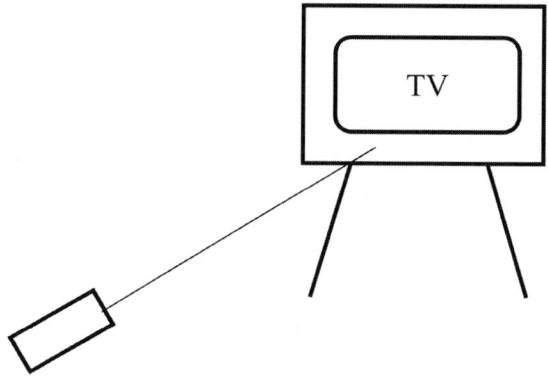

Figure 10.29   Principle 28. Mechanical interaction substitution. Use acoustic and optical interactions, taste and smell. Use electromagnetic fields.

■ See Chapter 9 for the discussion of the pattern of evolution called the increase of interactions. The history of technology is full of examples in which the mechanical means of doing something is first supplemented by an electrical system, then replaced by an electrical or electronic system. Automobile steering systems are mainly mechanical, but control by wire (already used extensively in aviation) is intensively studied in the automotive industry. In telecommunications, infrared and radio waves and other wireless technologies are increasingly used.

- In the case of automatic solar lawnmowers, something must prevent them from escaping to a neighbor's lawn. One solution is a sensor picking up a signal from a low-voltage (also solar-powered) cable buried out of sight.
- In communication and business, we also clearly see the increase of new interactions. When human society began, all communication was face-to-face, which has since been augmented by writing, telegraph, telephone, fax, e-mail, videoconferencing and other means.
- Transition to more easily controllable interactions is often associated with transition to the microlevel or segmentation. In inkjet printers, ink particles are controlled by thermal or electromagnetic fields. In video displays, text and figures are produced, changed and removed using electromagnetic fields to control micro-particles or molecules — for example, many flat-panel computer displays use liquid crystals, in which the image depends on the reflection of light from the molecules and the reflection is modified by changing the orientation of the molecule.

More examples:

- Maglev vehicles use magnetic fields to levitate above a guideway.
- Magnetic strip cards and smart cards are used instead of paper cash and checks.
- A ring laser gyroscope, unlike the old mechanical gyroscope, has no moving parts.
- CD devices with laser beam have superseded old record players with the diamond-tipped pickup arm.
- Have retail customers enter data by means of a touch screen, instead of filling out a form.

Exercise: Think of one example from your personal life or your business for each of the principles in this section.

10.11 PNEUMATICS AND HYDRAULICS, FLEXIBLE SHELLS, THIN FILMS AND POROUS MATERIALS (29–31)

10.11.1 Principle 29

Pneumatics and hydraulics. Use gas or liquid as parts of an object or system instead of solid parts (e.g., inflatable, filled with liquids, air cushion, hydrostatic, hydroreactive). See Figure 10.30.

25. Self-service	
26. Copying	
27. Cheap disposables	
28. Mechanical interaction substitution	

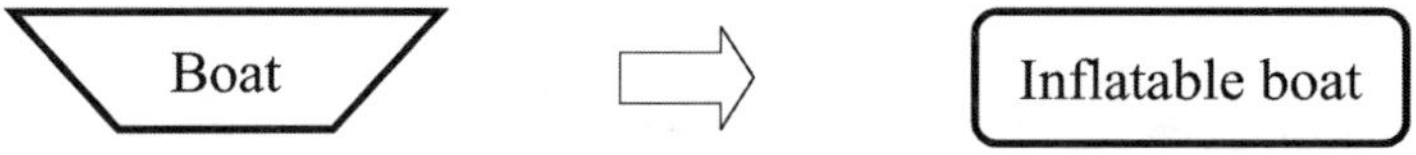

Boat ⟹ Inflatable boat

Figure 10.30   Principle 29. Pneumatics and hydraulics. Use gas and liquid instead of solid parts.

Examples:

- One way to transition to the microlevel is through the use of pneumatics and hydraulics. Examples of pneumatics are inflatable houses or air houses (stores, exhibition pavilions, sports halls and the like), inflatable boats and the moving of heavy components with an air cushion.
- Inflatable components allow the designer to decrease weight without losing strength. Air cushions enable movement of heavy objects using very little energy — for instance, they are used instead of mechanical jacks to raise aircraft that are on soft surfaces such as grass or mud.
- Similarly, hydraulic equipment can exert more force than simple mechanical systems of the same size. Replacing a physical cutting blade with a water jet illustrates the hydraulics principle of replacing solid objects with liquids.
- Business analogies use the idea of fluidity replacing rigid structures. Insurance companies now have "fluid" systems in which customers modify policies to suit their own needs. There is a lot of overlap with Principle 15 (dynamic parts).

■ In 1997, NASA used air bags, instead of expensive rockets, for the landing of Mars Pathfinder.

10.11.2 Principle 30

Flexible shells and thin films. Use flexible shells and thin films instead of three-dimensional structures. Isolate an object or system from the external environment using flexible shells and thin films (see Figure 10.31).

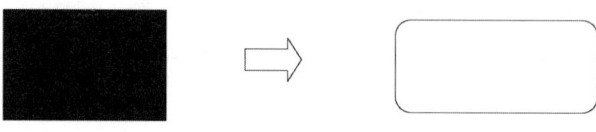

Figure 10.31   Principle 30. Flexible shells and thin films. Use flexible shells and thin films instead of three-dimensional structures.

Examples:

■ Coatings were historically spread on paper by means of a blade. One improvement is to transfer coating to the paper in the form of thin film. The paper is more evenly coated.

■ Medication in pill form is dispensed over a period of time by layering thin films that dissolve at different rates around small portions of the medicine. Patients can take one pill and the medicine will be released into their system over a period of time.

■ Some materials, such as epoxies, have two components that must be kept separate until shortly before use. Packages have premeasured portions that are separated by a thin film that the user can easily break to mix the product, making it convenient to use. This also has an element of Principle 10.

■ Heavy glass bottles for drinks are often replaced by cans made from thin metal (aluminum) or thin plastic material. Hydraulics is also used — the pressure of the drink makes the can stiff. Weight is reduced without deterioration of strength.

10.11.3 Principle 31

Porous materials. Make an object porous or add porous elements (inserts, coatings, etc.). If an object is already porous, use the pores to introduce a useful substance or function (see Figure 10.32).

Examples:

■ A ceramic filter made from hydrophilic porous sintered material simplifies and improves vacuum systems. Water that passes through

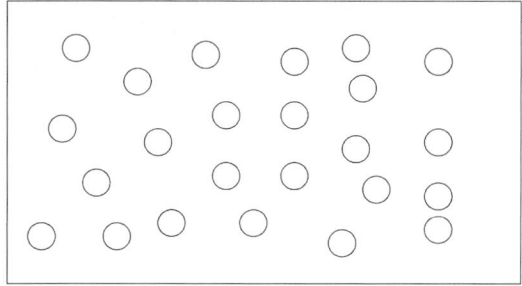

Figure 10.32   Principle 31. Porous materials. Make an object porous.

the filter seals the filter at the same time. The air leak of the conventional vacuum filter is eliminated and the consumption of energy decreased. Similarly, porous ceramics are used in many hazardous-waste-cleanup systems — the pores can be filled with materials that bind the hazardous substances and the ceramic particles make it much easier to distribute and remove these materials.

■ High-tech microfibers are now well known. Small pores prevent water from passing through, but allow moisture to evaporate. New plasters or wound dressings, are less known but not less exciting. They are covered by semi-permeable film. Water, dirt and bacteria are kept out, but excess moisture can evaporate through.

■ Business analogy: make the organization "porous" — make it easy for information to flow through the system. Make it easy for customers to penetrate the organization to find what they need.

Exercise: Think of one example from your personal life or your business for each of the principles in this section.

29. Pneumatics and hydraulics	
30. Flexible shells and thin films	
31. Porous materials	

10.12 OPTICAL PROPERTY CHANGES, HOMOGENEITY, DISCARDING AND RECOVERING (32–34)

10.12.1 Principle 32

Optical property changes. Change the color or transparency of an object or its external environment (see Figure 10.33).

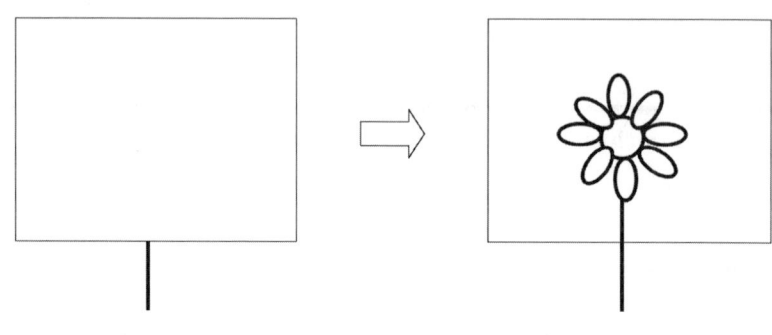

Figure 10.33 Principle 32. Optical property changes. Change the color or transparency of an object.

Examples:

- Enzymes producing light can be used for detecting impurities in food. Toys for preschool children are packed in transparent packages. Sunglasses that change the amount of light blocked, depending on the brightness of the environment, display both Principles 32 and 25 (self-service).
- Transparency is both a physical and a business term. Changing the transparency — increasing or decreasing it — is one important and often cheap way to improve business. Children want to see the toy; adults want to see that products and the whole production process are safe and ecological. Here transparency can make the difference between success and failure.
- Business school case studies often cite Johnson & Johnson for its exemplary "transparent" behavior as a form of crisis management. When there was a problem with tampering with its Tylenol product, the company immediately recalled the product and gave the public full information on its actions. Both the speed and the full disclosure of the situation are credited with the product's recovery and the company's retention of its excellent reputation.
- Porous materials and optical property changes compose a pair of principles that are frequently used together. Porous materials are

often semi-transparent. A business analogy is the firewall in a computer system. The wall should be transparent for legitimate users and should, at the same time, be impermeable to those who might try to steal essential information.

More examples:

- Use of lighting effects to change mood in a room or office
- Highlighter pens
- Creation of corporate colors — creating a strong brand image through use of bespoke colors — BP green, British Telecom red phone boxes, Ford blue, etc.
- "Tell the truth" tape turning a darker gray when food is spoiled

10.12.2 Principle 33

Homogeneity. Make objects that interact out of the same material (or material with identical properties). See Figure 10.34.

Figure 10.34 Principle 33. Homogeneity. Make objects that interact of the same material.

In the plastic industry, beads of polystyrene are frequently shipped in bags made of polystyrene. That way, the company that melts the beads can put the whole package, including the bags, into its processing equipment. This saves time — no need to open the bags — and eliminates the need to store and dispose of or recycle the bags.

In food, many products use the idea that the wrapper should have the same properties (be edible) as the contents. Ice cream cones, tacos and spring rolls are good examples.

The new wound dressings that we mentioned earlier act like a second skin. They keep wounds moist. Wounds themselves are moist and heal better in a moist environment. This innovation has an instructive history. From ancient Egypt to Rome and through the Middle Ages, wounds were treated by keeping them moist. Then this knowledge was forgotten and rediscovered only in the 1960s.

In business, this principle can be used by analogy. People may be more ready to buy things that remind them of familiar products than those that look very different. We are all familiar with movie sequels. *Godfather* was followed by *Godfather II* and *Godfather III* (and the Rambo series and the Jurassic Park series, etc.) Of course, the principle should be used together with the opposite idea. To make the movie sequel, you must first have an original movie, different from all others.

More examples:

- Co-located project teams
- Bioglass that bonds with natural bone
- Boeing working together teams — bringing customers and suppliers into the design loop
- Light bulbs replicating the balanced spectrum of more comfortable natural sunlight
- Instead of electric light, natural sunlight carried with light guides

10.12.3 Principle 34

Discarding and recovering. Make portions of an object that have fulfilled their functions go away (discard by dissolving, evaporating, etc.) or modify them directly during operation. Conversely, restore consumable parts of an object directly in operation (see Figure 10.35).

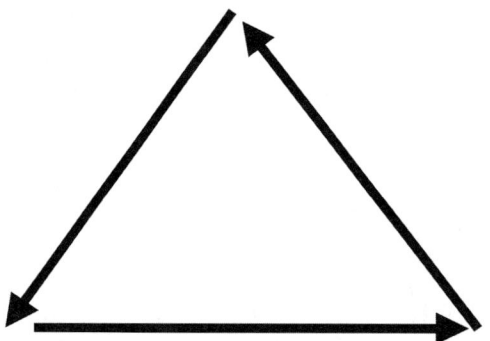

Figure 10.35   Principle 34. Discarding and recovering. Make portions of an object that have fulfilled their functions go away. Restore consumable parts.

Examples:

- Biodegradable materials in medicine: polylactides are used to make dissolvable screws and pins. They can replace titanium screws used by surgeons to mend broken bones. The second operation for the removal of screws is not needed. Another example: the tread of a tire that restores the edges of tread blocks when the tire wears.

■ Many recovery processes have been used in technology for a long time. Chemicals used for pulping wood are recovered in a recovery boiler. Environmental requirements will make recovery more popular. For example, water polluted in some process is more and more often purified and recycled back to the process.

■ In business, the project organization is a good example of discarding and recovering. A good project should have an end. The organization will then be dissolved. The members can use their skills again in new projects. In all work, knowledge and skills are updated and improved directly during work and by retraining.

Exercise: Think of one example from your personal life or your business for each of the principles in this section.

32. Optical property changes	
33. Homogeneity	
34. Discarding and recovering	

10.13 PARAMETER CHANGES, PHASE TRANSITIONS AND THERMAL EXPANSION (35–37)

10.13.1 Principle 35

Parameter changes. Change an object's physical state (e.g., to a gas, liquid or solid). Change the concentration or consistency. Change the degree of flexibility. Change the temperature (see Figure 10.36).

Examples:

■ The new medical plaster described earlier also illustrates a parameter change. A traditional dry plaster is replaced by a moist one.

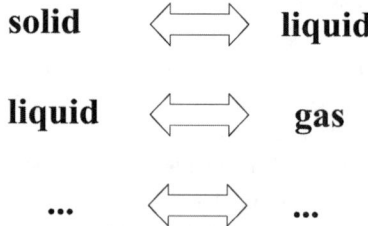

Figure 10.36   Principle 35. Parameter changes. Change an object's physical state.

The principles of parameter change, homogeneity and use of porous materials are combined in this technology.

■ Transport gaseous nitrogen or oxygen as liquids; similarly, transport natural gas and propane as liquids.

■ Change the melting temperature of chocolate so that it doesn't melt when carried in high temperatures.

■ Powdered paints can be used instead of liquid ones. Powder paint combines the performance of modern latex and silicon-based emulsions with the convenience of the powder. Powder can be easily transported and stored. To use, just add water.

■ In business situations, a parameter change is frequently realized as a policy change. In the past decade, many companies have increased the flexibility of employee benefit programs — instead of having one standard program, employees can design a mix of medical and life insurance, pension plans, etc. Likewise, mass customization systems let customers have much more flexibility in designing products that exactly fit their needs (also a demonstration of Principle 15, dynamic parts.)

10.13.2 Principle 36

Phase transitions. Use phenomena occurring during phase transitions (e.g., volume changes, loss or absorption of heat, etc.). The most common of the many kinds of phase transitions include solid–liquid–gas–plasma, paramagnetic–ferromagnetic and normal conductor–superconductor, but many useful phenomena are associated with more exotic transitions as well, such as solid–solid crystallographic changes, superfluidity, antiferromagnetism, etc. (see Figure 10.37).

Examples:

■ Blasting with solid carbon dioxide can clean surfaces. Impurities will freeze immediately, contract and loosen easily. The example

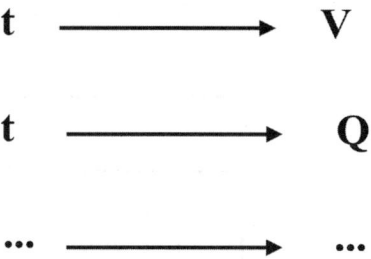

Figure 10.37   Principle 36. Phase transitions. Use phenomena occurring during phase transitions.

also illustrates parameter changes (35), thermal expansion (37) and hurrying (21). Impurities are frozen so quickly that the cleaned material doesn't suffer from thermal expansion. The carbon dioxide then sublimes (harmlessly turns into gas), so there is no cleanup needed.

- Heat pipes are well known examples of using the phenomena associated with phase transitions. The heat given up and absorbed as the fluid in the pipe transitions from liquid to gas can be used either as a heater or air conditioner, depending on how the system is arranged.
- "Muscle wire" is a form of nickel-titanium alloy that is used in robot systems and in orthodontics. Small electric currents heat the wire, which goes through a crystallographic phase change that changes its length.
- When businesses make structural changes (mergers, acquisitions or internal changes) the accompanying phenomena are analogous to heat in a phase change — there is lots of confusion. Constructive ways to use this period of disruption include finding new means of aligning business systems with new strategies, forming new alliances with customers or suppliers and getting rid of obsolete practices.

More examples:

- Thermal inkjet uses vaporization to force ink out of the nozzle.
- Be aware of the requirements of different stages — conception, birth, development, maturity, retirement — of a project.

10.13.3 Principle 37

Thermal expansion. Use thermal expansion (or contraction) of materials. If thermal expansion is being used, choose multiple materials with different coefficients of thermal expansion (see Figure 10.38).

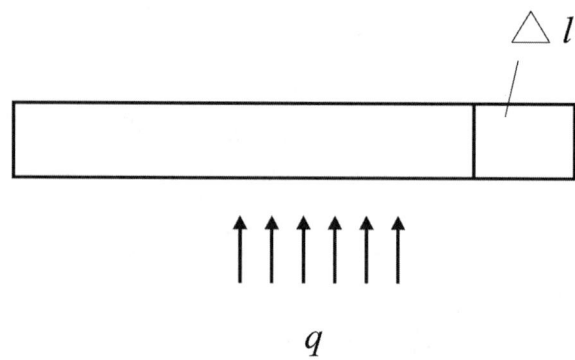

Figure 10.38   Principle 37. Thermal expansion. Use thermal expansion.

Thermal expansion can be used to position and fit components such as a valve in an engine. A component is cooled in liquid nitrogen, contracts, is installed, then expands and fixes itself in position. A kitchen example is loosening a tight metal lid on a glass jar by running it under hot water. The metal lid expands more than the glass jar, so it is easier to remove the lid.

Exercise: Think of one example from your personal life or your business for each of the principles in this section.

35. Parameter changes	
36. Phase transitions	
37. Thermal expansion	

10.14 STRONG OXIDANTS, INERT ATMOSPHERE AND COMPOSITE MATERIALS (38–40)

10.14.1   Principle 38

Strong oxidants. Replace common air with oxygen-enriched air. Replace enriched air with pure oxygen. Expose air or oxygen to ionizing radiation.

Use ionized oxygen. Replace ozonized air (ionized oxygen) with ozone (see Figure 10.39). Oxygen is used in the bleaching of pulp (for paper production). There are ideas and experiments to use ozone for bleaching.

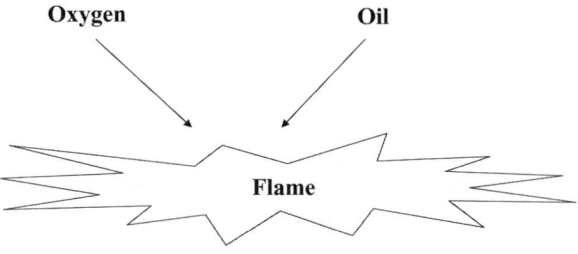

Figure 10.39   Principle 38. Strong oxidants. Make the process more active.

Jules Verne wrote a novel called *Dr. Ox's Experiment*. In Quiquendone, an imaginary town, life is very slow. Then Dr. Ox injected a gas into the town and everything started happening very fast. People got exceptionally lively and passionate. Perhaps we also sometimes need something like Dr. Ox's gas to charge the mental atmosphere with positive effects.

Examples:

- Using simulations and games instead of lecture-style training
- Scuba diving with Nitrox or other non-air mixtures to extend endurance
- Treating wounds in a higher pressure oxygen environment to kill anaerobic bacteria
- Metal active gas welding (MAG)

10.14.2   Principle 39

Inert atmosphere. Replace a normal environment with an inert one. Add neutral parts or inert additives to an object or system (see Figure 10.40).

Examples:

- Inert gases (such as carbon dioxide or argon) are used in welding to prevent oxidation of the material at the weld.
- Inert materials are added to detergents to make them easier for consumers to measure. (Did you ever wonder why the box says "97% inert materials"?)

Strong oxidants and inert atmosphere can be considered as a pair of principles. Using them together frequently gives good results. Oxygen is

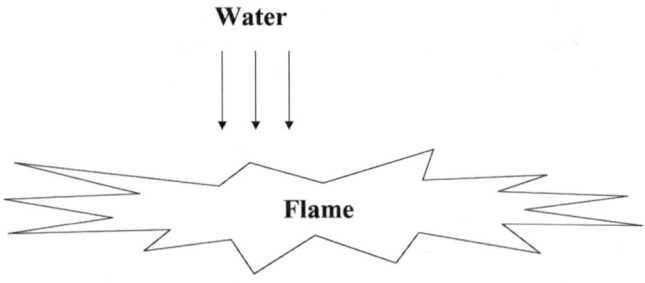

Figure 10.40 Principle 39. Inert atmosphere. Make process slower and more passive.

used to generate needed energy and inert gases are used to prevent undesired oxidation. An inventor proposed using a welding device as a fire extinguisher. A device is provided with a simple system that can blast inert gas with high pressure.

A social analogy of inert atmosphere may be indifference and neutrality. In business, negative situations mostly need cooling. Ignore or neutralize negative and destructive actions (if you cannot turn them positive; see Principle 22). Use neutral arbitrators. Here, too, the point is to use the correct approach at the right time.

Another example:

■ Metal inert gas welding (MIG)

10.14.3 Principle 40

Composite materials. Change from uniform to composite (multiple) materials and systems (see Figure 10.41).

Examples:

■ Rubber reinforced with woven cords, reinforced concrete and glass-fiber-reinforced plastics are typical technology examples.
■ The use of *nothing* (air or vacuum) as one of the elements of a composite is very typical of TRIZ — no resource is an available in all situations. Examples include honeycomb materials (egg crates, aircraft structures), hollow systems (golf clubs, bones) and sponge materials (packaging materials, scuba diving suits.) These combine Principle 31, use of porous material, with Principle 40, use of composite materials.
■ In business, we can speak of composite structures as well. Multi-disciplinary project teams are often more effective than groups representing experts from one field. Multimedia presentations often do better in marketing, teaching, learning and entertainment than

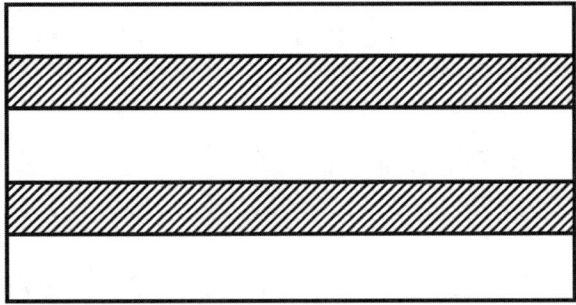

Figure 10.41   Principle 40. Composite materials. Change from uniform to composite systems.

single-medium performances. Other examples are less tangible but not at all less important. Fanatic commitment to cleanliness is one famous feature of McDonald's. Consistent preparation of food is another major commitment. These are two principles or values or fibers that tie together a loose organization.

The principle of composite materials or, more generally, composite systems, is a good conclusion for this section on using inventive principles. If you have a system, you can improve the result by combining it with another system. Innovative principles are also systems. Composite principles often do better than single ones.

More examples:

- Combined high risk/low risk investment strategy
- Flammable polyurethane coated with fire-resistant Kevlar, e.g., in airplane seat cushions

Exercise: Think of one example from your personal life or your business for each of the principles in this section.

38.Strong oxidants	
39.Inert atmosphere	
40.Composite materials	

You can make the 40 principles a more effective toolkit and tailor them to your needs and purposes by adding your own examples. Some people find they get the most creative stimulus from examples outside their own field and others prefer to start with examples that are close to their own field. If you share your examples with others, you will contribute to the continuous development of problem-solving tools.

10.15 HOW TO SELECT THE MOST SUITABLE PRINCIPLES

The easiest way to find a useful principle is simply to browse the list in a relaxed way. Knowledge of the system and the constraints in the problem will help you decide which principles are suitable.

Table 10.1   Examples of Problems and Principles that Can Be Used to Resolve Them

Problems	Principles
Reducing lawnmower noise	Porous materials (31)
Making it easy to grow carrots	Merging (5), multifunctionality (6)
Improving pin-type latch	Segmentation (1), dynamic parts (15)
Making training more effective	Segmentation (1), composite systems (40)
Firefighting with less water	Segmentation (1)

Let's consider five earlier examples: reducing lawnmower noise, making it easy to grow carrots, improving the pin-type latch, making training more effective and fighting fire with less water. It is easy to see the applicability of certain principles to each of the examples.

Table 10.1 reflects typical results obtained by quickly browsing the list of principles. Closer examination reveals more principles — perhaps more suitable ones. These results can be improved using some guidelines for the selection of the most helpful principles.

10.15.1 Tradeoffs

Principles can be selected by tradeoffs (considered in Chapter 3). When the tradeoff is formulated, it can be used to eliminate unsuitable principles. Principles that negate the tradeoff are those that should be selected. For example, when noise absorption improves, the size of the muffler and the number of parts increase. Use the multifunctionality or universality principle (6). If the increasing size and part number are the problem, it is wise to check whether some other resource, for example the casing or the grass, can take the job of the muffler.

The more precisely carrot seeds are planted, the slower the speed of planting. There is a process — planting. Second, there are many seeds, which are difficult to handle. Preliminary action (10) and the intermediary principle (24) seem natural. Both help remove the contradiction by suggesting that the carrot seeds could be arranged on the tape before planting.

If a pin is made easy to lock and open, wear increases. The same component should meet contradictory requirements and have incompatible properties. The segmentation principle (1) may be the first approximation, because the parts and the whole can have opposite properties. For example, if the pin is segmented to two parts so that one part is a small wedge, the latch may be easier to open. Use of asymmetry (4) will also help — the pin can have different dimensions at the top, middle and bottom because each area has different requirements.

As training gets more thorough, it requires more time. These are contradictory requirements for the same activity. Segmentation (1), multifunctionality (6) preliminary action (10) and composite materials (40) may be the first principles that come to mind. Lengthy training can be broken down into smaller parts that can be incorporated into work and perhaps prepared beforehand.

The more water that is used for fighting fires, the more equipment is needed and the more damage will occur. Consider segmentation (1). Again, the same component should meet contradictory requirements and have incompatible properties. First we can try to fragment water some way.

Additional cues for finding principles for resolving tradeoffs are summarized in the contradiction matrix. The matrix is a big table. The rows are labeled with typical features that should be improved and the columns are labeled with the features that get worse. The cell at the intersection of the row and column has the numbers of the principles that were historically used most frequently to solve this particular tradeoff (see Table 10.2).

Typical features of the matrix are presented in Table 10.3 (and Reference 8) and a modification of the full matrix, also called Altshuller's matrix, which appears as an appendix at the end of the chapter. Altshuller developed it first, together with principles of innovation. The original matrix is also available on the Internet.[9] In our version, numbers in the cells are the same, but cells that were empty now have the word "all" to indicate that all 40 principles should be tried.

To work with the matrix, follow these steps:

1. Select the standard feature from Table 10.3 nearest to the feature that should be improved.
2. Select the standard feature from Table 10.3 nearest to the feature that will worsen in this case.
3. Find the row in the matrix (Appendix) with the number of the improving standard feature.
4. Find the column in the Appendix with the number of the worsening standard feature.
5. In the cell at the intersection of the row and column are the numbers of recommended principles.
6. Look up those principles in the list of 40 principles and use them to create ideas for solving your problem. Don't be surprised if some of the recommended principles don't help or if you find great ideas from principles that were not recommended — the numbers in the cells are based on historical statistics, not on any knowledge of your particular problem.

10.15.2 Inherent Contradictions and Resources

Principles can be selected by inherent contradictions and by resources (considered in Chapters 4 and 5). The constraints of the problem may require that opposite requirements are satisfied at different times, in different places or at the same time at the same place. The three upper rows in Table 10.4 present these situations.

Let's try these recommendations on some of the examples we have been working with.

Big muffler–no muffler. Because the muffler should be big and have zero size at the same time in the same place, multifunctionality principle (6) is the first to think of.

Many seeds–single seed. Because there should be many seeds and individual seeds at the same time in the same place, the merging principle (5) may be the first principle to try.

Big clearance–no clearance. Because the clearance should be big and zero at different times, Segmentation (1) and dynamic parts (15) can be selected. Because the pin and the hole have simple geometric forms, geometric principles are also interesting.

Long training time–no training time. Training should be long and short at the same time in the same place. Segmentation (1) and multifunctionality (6) are good principles.

Table 10.2 Selected Rows and Columns from the Contradiction Matrix

Improved Feature	Worsened Feature								
	31	32	33	34	35	36	37	38	39
11 Stress or pressure	2, 33, 27, 18	1, 35, 16	11	2	35	19, 1, 35	2, 36, 37	35, 24	10, 14, 35, 37
12 Shape	35, 1	1, 32, 17, 28	32, 15, 26	2, 13, 1	1, 15, 29	16, 29, 1, 28	15, 13, 39	15, 1, 32	17, 26, 34, 10
13 Stability of the object's composition	35, 40, 27, 39	35, 19	32, 35, 30	2, 35, 10, 16	35, 30, 34, 2	2, 35, 22, 26	35, 22, 39, 23	1, 8, 35	23, 35, 40, 3
14 Strength	15, 35, 22, 2	11, 3, 10, 32	32, 40, 25, 2	27, 11, 3	15, 3, 32	2, 13, 25, 28	27, 3, 15, 40	15	29, 35, 10, 14
15 Duration of action by a moving object	**21, 39, 16, 22**	27, 1, 4	12, 27	29, 10, 27	1, 35, 13	10, 4, 29, 15	19, 29, 39, 35	6, 10	35, 17, 14, 19

Note: The numbers in the cell refer to the principles that have the highest probability (based on historical analysis) of resolving the contradiction. For example, consider the proposal to change the speed of inflation of an airbag to reduce injuries to small occupants. The tradeoff is that injuries in high-speed accidents increase. Translating this into the TRIZ matrix terms, the parameter that improves is "Duration of action of a moving object" (Row 15) and the parameter that worsens is "Object-generated harmful effects" (Column 31). The cell at the intersection has the notation "21, 39, 16, 22." These are the identifiers for four of the principles of invention.

Table 10.3   Explanation of the 39 Features of the Contradiction Matrix

	Title	*Explanation*
	Moving objects	Objects that can easily change position in space, either on their own or as a result of external forces Vehicles and objects designed to be portable are the basic members of this class
	Stationary objects	Objects that do not change position in space, either on their own or as a result of external forces Consider the conditions under which the object is being used
1	Weight of moving object	Mass of the object in a gravitational field Force that the body exerts on its support or suspension
2	Weight of stationary object	Mass of the object in a gravitational field Force the body exerts on its support or suspension or on the surface on which it rests
3	Length of moving object	Any one linear dimension, not necessarily the longest, is considered a length
4	Length of stationary object	Same
5	Area of moving object	A geometrical characteristic described by the part of a plane enclosed by a line Part of a surface occupied by the object OR the square measure of the surface, either internal or external, of an object
6	Area of stationary object	Same Length × width × height for a rectangular object, height × area for a cylinder, etc.
7	Volume of moving object	Cubic measure of space occupied by the object
8	Volume of stationary object	Same
9	Speed	Velocity of an object Rate of a process or action in time
10	Force	Force measures the interaction between systems. In Newtonian physics, force = mass × acceleration. In TRIZ, force is any interaction that is intended to change an object's condition.
11	Stress or pressure	Force per unit area Tension
12	Shape	External contours, appearance of a system

Table 10.3 (Continued)   Explanation of the 39 Features of the Contradiction Matrix

	Title	*Explanation*
13	Stability of object's composition	Wholeness or integrity of the system Relationship of system's constituent elements Wear, chemical decomposition and disassembly are all decreases in stability. Increasing entropy is decreasing stability
14	Strength	Extent to which the object is able to resist changing in response to force Resistance to breaking
15	Duration of action by a moving object	Time that the object can perform the action Service life Mean time between failure is a measure of the duration of action
		Durability
16	Duration of action by stationary object	Same
17	Temperature	Thermal condition of the object or system Loosely includes other thermal parameters, such as heat capacity, that affect the rate of change of temperature
18	Illumination intensity	Light flux per unit area, also any other illumination characteristics of the system such as brightness, light quality, etc.
19	Use of energy by moving object	Measure of the object's capacity for doing work In classical mechanics, energy is the product of force × distance. This includes the use of energy provided by the super system (such as electrical energy or heat.) Energy required to do a particular job
20	Use of energy by stationary object	Same
21	Power	Time rate at which work is performed Rate of use of energy
22	Loss of energy	Use of energy that does not contribute to the job being done (see 19)
		Reducing the loss of energy sometimes requires techniques that differ from improving the use of energy, which is why this is a separate category

Table 10.3 (Continued)   Explanation of the 39 Features of the Contradiction Matrix

	Title	Explanation
23	Loss of substance	Partial or complete, permanent or temporary, loss of some of a system's materials, substances, parts or subsystems
24	Loss of information	Partial or complete, permanent or temporary, loss of data or access to data in or by a system. Frequently includes sensory data such as aroma, texture, etc.
25	Loss of time	Time is the duration of an activity. Improving the loss of time means reducing the time taken for the activity. "Cycle time reduction" is a common term
26	Quantity of substance /quantity of matter	The number or amount of a system's materials, substances, parts or subsystems that might be changed fully or partially, permanently or temporarily
27	Reliability	A system's ability to perform its intended functions in predictable ways and conditions
28	Measurement accuracy	Closeness of the measured value to the actual value of a property of a system
		Reducing the error in a measurement increases the accuracy of the measurement
29	Manufacturing precision	Extent to which the actual characteristics of the system or object match the specified or required characteristics
30	External harm affects the object	Susceptibility of a system to externally generated (harmful) effects
31	Object-generated harmful factors	A harmful effect reduces the efficiency or quality of the functioning of the object or system, generated by the object or system as part of its operation
32	Ease of manufacture	Degree of facility, comfort or effortlessness in manufacturing or fabricating object or system
33	Ease of operation	Simplicity: The process is NOT easy if it requires many people, many steps in the operation, needs special tools, etc. "Hard" processes = low yield; "easy" processes = high yield; they are easy to do right
34	Ease of repair	Quality characteristics such as convenience, comfort, simplicity and time to repair faults, failures or defects in a system

Table 10.3 (Continued)   Explanation of the 39 Features of the Contradiction Matrix

	Title	Explanation
35	Adaptability or versatility	The extent to which a system or object responds positively to external changes A system that can be used in multiple ways in a variety of circumstances
36	Device complexity	Number and diversity of elements and element interrelationships within a system. User may be an element of the system that increases the complexity. The difficulty of mastering the system is a measure of its complexity
37	Difficulty of detecting and measuring	Measuring or monitoring systems that are complex and costly, require much time and labor to set up and use or that have complex relationships between components or components that interfere with each other all demonstrate "difficulty of detecting and measuring." Increasing cost of measuring to a satisfactory error is also a sign of increased difficulty of measuring
38	Extent of automation	The extent to which a system or object performs its functions without human interface. The lowest level of automation is the use of a manually operated tool. For intermediate levels, humans program the tool, observe its operation and interrupt or reprogram as needed. For the highest level, the machine senses the operation needed, programs itself and monitors its own operations.
39	Productivity	The number of functions or operations performed by a system per unit time The time for a unit function or operation The output per unit time or the cost per unit output

Much water–no water. There should be much water to suppress the fire and there should be no water, to keep the equipment simple and to decrease water damage. Principle: segmentation (1).

Depending on the system, a variety of general resources such as air, water, space and gravity are available. The last row of Table 10.4 shows some principles that can help you to use these resources.

Always consider the components of the resource and modifications of the resource. If you have air, you have nitrogen, oxygen, carbon dioxide and small concentrations of other gases. If you have sunlight, you have yellow light, red light, ultraviolet light, infrared light, etc. If you have DC electric power, you can make a loop in the wire carrying the electricity and make a magnet. If you have water and power, you could make steam or ice. Then, apply the principles to these new resources.

Table 10.4   Some Principles Related to Certain Contradictions and Resources

Contradiction and Resource	Principles
Principles for realizing incompatible requirements in different times (time resources)	Segmentation (1), preliminary counteraction (9), preliminary action (10), beforehand compensation (11), dynamic parts (15), periodic action (19), hurrying (21), discarding and recovering (34)
Principles for realizing incompatible requirements in different places (space resources)	Separation (2), local quality (3), symmetry change (4), "nested doll" (7), dimensionality change (17), flexible shells and thin films (30)
Principles for realizing incompatible requirements at the same time in the same place (resources on the macro- and microlevel)	Segmentation (1), merging (5), multifunctionality (6), intermediary (24), composite materials (40)
General resources: air, water, space, gravity, others	Pneumatics and hydraulics (29), weight compensation (8), inert atmosphere (39)

10.15.3 Using the Features of the Ideal Final Result

Principles can be selected by the features of the ideal final result (considered in Chapter 6).

The ideal system has no weight, no size and consumes no energy. Principles connected most directly with the increase of ideality are shown on the last row of Table 10.5. Some other component should deliver the function of the muffler or the pin. Some other activity, for example work, should include training. Seeding should be combined with other operations. Training should become self-training, etc.

Table 10.5   Patterns of Evolution and Related Principles

Pattern	Principles
Uneven evolution of the system	All
Transition to macrolevel	Merging (5). "nested doll" (7), weight compensation (8), multifunctionality (6), composite materials (40)
Transition to microlevel	Segmentation (1), "nested doll" (7), mechanical interaction substitution (28), flexible shells and thin films (30), porous materials (31), parameter change (35), phase transition (36)
The increase of interactions: introduction of substances and actions	Porous materials (Principle 31), inert atmosphere (39), feedback (23), intermediary (24), mechanical interaction substitution (28), thermal expansion (37), parameter change (35)
Expansion and convolution	Separation (2) Local quality (3), merging (5), multi-functionality (6)
Summary: increasing of ideality	Multi-functionality (6), "Blessing in disguise" (22), self-service (25)

It will be particularly helpful to formulate the ideal final result and why the ideal cannot be achieved. This will usually lead to contradictions. For example,[10] in a customer service situation, the ideal final result is that customers get exactly what they need exactly when they need it. An analysis of the problem might follow this path:

> "I can't give it to her because my employees don't have all the knowledge."

> "Why not?"

> "Because employee turnover is so fast that untrained employees are used."

This analysis reveals several potential problems and families of solutions:

- Customers get what they need without the (direct) help of employees.
- Employees have the knowledge without training.
- Trained employees don't leave the job.

Now use the 40 principles to look for solutions to each of these categories of problems, then select the one (or more) that has the highest probability of working in this situation. When applying the 40 inventive principles, keep in mind the TRIZ concepts of removing the reason for the contradiction and using available resources.

Principles can be selected by the patterns of evolution (considered in Chapter 9). Some principles are the same as the patterns and some others can be easily seen as ways to use patterns. Conversely, the patterns can be seen as arising from the repeated use of certain sets of principles. Table 10.5 illustrates the correspondence between patterns and principles: only the most obvious links are shown in the table. The more you use the principles, the more you will see connections between them and evolution patterns.

You might make different choices of principles to apply to the example problems. The examples shown here are not at all exhaustive. The principles are good ideas, based on many strong solutions to a wide variety of problems over many years. Which solutions will be best for your particular problem will depend on the problem, the resources and the constraints. Try using the 40 principles to get new insights that can help you find new ways to approach solving your problem (Table 10.6) .

Table 10.6 Apply Innovative Principles to Your Own Problem

What is the contradiction?

If this contradiction is in the Contradiction Matrix, which principles are suggested?

If this is an inherent contradiction, which principles might help?

What ideas did you get from each principle?

Browse through all the principles. What additional ideas did you get?

10.16 SUMMARY

- Innovative principles give clues for solving problems. The simplest way to use the principles is to go through them all.
- Some selection criteria are useful for deciding which principles to use for any particular problem. Seek principles that resolve tradeoffs. Try the Contradiction Matrix if it fits your tradeoffs.[9] Seek principles that eliminate the inherent contradictions. Resource map-

ping, the description of the ideal final result and the patterns of evolution also help to find useful principles.

ACKNOWLEDGMENTS

The authors would like to thank Darrell Mann and John Terninko for generously sharing their lists of examples of many of the 40 principles. We also thank Joe Miller and Ellen MacGran for their help with the research for features of the matrix.

REFERENCES

1. Altshuller, G.S., *40 Principles*, TIC, Worcester, 1997.
2. Savransky, S.D. , *Engineering of Creativity*, CRC, Boca Raton, 2000, 199.
3. Zlotin, B. and Zusman, A., Managing Innovation Knowledge, *Izobretenie*, I, 21, 1999.
4. E. Domb and Altshuller, G.S., *40 Principles*, TIC, Worcester, 1997, 125.
5. Mann, D., Using TRIZ to Overcome The Mass Customization Contradiction, presented at TRIZ conference 'TRIZ Future 2001', Bath, November 7-9, 2001.
6. Mann, D. and Domb, E., Business Contradictions — 1) Mass Customization, *TRIZ Journal*, December 1999, www.triz-journal.com.
7. *How Things Work Today*, Wright, M. and Patel, M., Eds., Marshall Publishing, London, 2000, 303.
8. Domb, E. The 39 features of Altshuller's Contradiction Matrix, *Triz Journal*, November, 1998, www.triz-journal.com.
9. Contradiction Matrix, *TRIZ Journal*, July 1997, www.triz-journal.com.
10. Mann, D. and Domb, E. 4 inventive (business) principles with examples, *Triz Journal*, September, 1999, www.triz-journal.com.

Appendix   The Contradiction Matrix

Improved Feature	Worsened Feature									
	1	2	3	4	5	6	7	8	9	10
1 Weight of moving object	all	all	15, 8, 29,34	all	29, 17, 38, 34	all	29, 2, 40, 28	all	2, 8,15, 38	8, 10, 18, 37
2 Weight of stationary object	all	all	all	10, 1, 29, 35	all	35, 30, 13, 2	all	5, 35, 14, 2	all	8, 10, 19, 35
3 Length of moving object	8, 15, 29, 34	all	all	all	15, 17, 4	all	7, 17, 4, 35	all	13, 4, 8	17, 10, 4
4 Length of stationary object	all	35, 28, 40, 29	all	all	all	17, 7, 10, 40	all	35, 8, 2, 14	all	28, 10
5 Area of moving object	2, 17, 29, 4	all	14, 15, 18, 4	all	all	all	7, 14, 17, 4	all	29, 30, 4, 34	19, 30, 35, 2
6 Area of stationary object	all	30, 2, 14, 18	all	26,7, 9, 39	all	all	all	all	all	1, 18, 35, 36
7 Volume of moving object	2, 26, 29, 40	all	1, 7, 4, 35	all	1, 7, 4, 17	all	all	all	29, 4, 38, 34	15, 35, 36, 37
8 Volume of stationary object	all	35, 10, 19, 14	19, 14	35, 8, 2, 14	all	all	all	all	all	2, 18, 37
9 Speed	2, 28, 13, 38	all	13, 14, 8	all	29, 30, 34	all	7, 29, 34	all	all	13, 28, 15, 19
10 Force	8,1,37, 18	18, 13, 1, 28	17, 19, 9, 36	28, 10	19, 10, 15	1, 18, 36, 37	15, 9, 12, 37	2, 36, 18, 37	13, 28, 15, 12	all

11	Stress or pressure	10, 36, 37, 40	13, 29, 10, 18	35, 10, 36	35, 1, 14, 16	10, 15, 36, 28	10, 15, 36, 37	6, 35, 10	35, 24	6, 35, 36	36, 35, 21
12	Shape	8, 10, 29, 40	15, 10, 26, 3	29, 34, 5, 4	13, 14, 10, 7	5, 34, 4, 10	all	14, 4, 15, 22	7, 2, 35	35, 15, 34, 18	35, 10, 37, 40
13	Stability of the object's composition	21, 35, 2, 39	26, 39, 1, 40	13, 15, 1, 28	37	2, 11, 13	39	28, 10, 19, 39	34, 28, 35, 40	33, 15, 28, 18	10, 35, 21, 16
14	Strength	1, 8, 40, 15	40, 26, 27, 1	1, 15, 8, 35	15, 14, 28, 26	3, 34, 40, 29	9, 40, 28	10, 15, 14, 7	9, 14, 17, 15	8, 13, 26, 14	10, 18, 3, 14
15	Duration of action by a moving object	19, 5, 34, 31	all	2, 19, 9	all	3, 17, 19	all	10, 2, 19, 30	all	3, 35, 5	19, 2, 16
16	Duration of action by stationary object	all	6, 27, 19, 16	all	1, 40, 35	all	all	all	35, 34, 38	all	all
17	Temperature	36, 22, 6, 38	22, 35, 32	15, 19, 9	15, 19, 9	3, 35, 39, 18	35, 38	34, 39, 40, 18	35, 6, 4	2, 28, 36, 30	35, 10, 3, 21
18	Illumination intensity	19, 1, 32	2, 35, 32	19, 32, 16	all	19, 32, 26	all	2, 13, 10	all	10, 13, 19	26, 19, 6
19	Use of energy by moving object	12, 18, 28, 31	all	12, 28	all	15, 19, 25	all	35, 13, 18	all	8, 35, 35	16, 26, 21, 2
20	Use of energy by stationary object	all	19, 9, 6, 27	all	all	all	all	all	all	all	36, 37

(continued)

Appendix (CONTINUED) The Contradiction Matrix

		Improved Feature	1	2	3	4	5	6	7	8	9	10
								Worsened Feature				
21	Power		8, 36, 38, 31	19, 26, 17, 27	1, 10, 35, 37		19, 38	17, 32, 13, 38	35, 6, 38	30, 6, 25	15, 35, 2	26, 2, 36, 35
22	Loss of Energy		15, 6, 19, 28	19, 6, 18, 9	7, 2, 6, 13	6, 38, 7	15, 26, 17, 30	17, 7, 30, 18	7, 18, 23	7	16, 35, 38	36, 38
23	Loss of substance		35, 6, 23, 40	35, 6, 22, 32	14, 29, 10, 39	10, 28, 24	35, 2, 10, 31	10, 18, 39, 31	1, 29, 30, 36	3, 39, 18, 31	10, 13, 28, 38	14, 15, 18, 40
24	Loss of Information		10, 24, 35	10, 35, 5	1, 26	26	30, 26	30, 16		2, 22	26, 32	
25	Loss of Time		10, 20, 37, 35	10, 20, 26, 5	15, 2, 29	30, 24, 14, 5	26, 4, 5, 16	10, 35, 17, 4	2, 5, 34, 10	35, 16, 32, 18	all	10, 37, 36, 5
26	Quantity of substance/the matter		35, 6, 18, 31	27, 26, 18, 35	29, 14, 35, 18	all	15, 14, 29	2, 18, 40, 4	15, 20, 29	all	35, 29, 34, 28	35, 14, 3
27	Reliability		3, 8, 10, 40	3, 10, 8, 28	15, 9, 14, 4	15, 29, 28, 11	17, 10, 14, 16	32, 35, 40, 4	3, 10, 14, 24	2, 35, 24	21, 35, 11, 28	8, 28, 10, 3
28	Measurement accuracy		32, 35, 26, 28	28, 35, 25, 26	28, 26, 5, 16	32, 28, 3, 16	26, 28, 32, 3	26, 28, 32, 3	32, 13, 6	all	28, 13, 32, 24	32, 2
29	Manufacturing precision		28, 32, 13, 18	28, 35, 27, 9	10, 28, 29, 37	2, 32, 10	28, 33, 29, 32	2, 29, 18, 36	32, 23, 2	25, 10, 35	10, 28, 32	28, 19, 34, 36
30	External harm affects the object		22, 21, 27, 39	2, 22, 13, 24	17, 1, 39, 4	1, 18	22, 1, 33, 28	27, 2, 39, 35	22, 23, 37, 35	34, 39, 19, 27	21, 22, 35, 28	13, 35, 39, 18

31	Object-generated harmful factors	19, 22, 15, 39	35, 22, 1, 39	17, 15, 16, 22	all	17, 2, 18, 39	22, 1, 40	17, 2, 40	30, 18, 35, 4	35, 28, 3, 23	35, 28, 1, 40
32	Ease of manufacture	28, 29, 15, 16	1, 27, 36, 13	1, 29, 13, 17	15, 17, 27	13, 1, 26, 12	16, 40	13, 29, 1, 40	35	35, 13, 8, 1	35, 12
33	Ease of operation	25, 2, 13, 15	6, 13, 1, 25	1, 17, 13, 12	all	1, 17, 13, 16	18, 16, 15, 39	1, 16, 35, 15	4, 18, 39, 31	18, 13, 34	28, 13, 35
34	Ease of repair	2, 27, 35, 11	2, 27, 35, 11	1, 28, 10, 25	3, 18, 31	15, 13, 32	16, 25	25, 2, 35, 11	1	34, 9	1, 11, 10
35	Adaptability or versatility	1, 6, 15, 8	19, 15, 29, 16	35, 1, 29, 2	1, 35, 16	35, 30, 29, 7	15, 16	15, 35, 29	all	35, 10, 14	15, 17, 20
36	Device complexity	26, 30, 34, 36	2, 26, 35, 39	1, 19, 26, 24	26	14, 1, 13, 16	6, 36	34, 26, 6	1, 16	34, 10, 28	26, 16
37	Difficulty of detecting and measuring	27, 26, 28, 13	6, 13, 28, 1	16, 17, 26, 24	26	2, 13, 18, 17	2, 39, 30, 16	29, 1, 4, 16	2, 18, 26, 31	3, 4, 16, 35	30, 28, 40, 19
38	Extent of automation	28, 26, 18, 35	28, 26, 35, 10	14, 13, 17, 28	23	17, 14, 13	all	35, 13, 16	all	28, 10	2, 35
39	Productivity	35, 26, 24, 37	28, 27, 15, 3	18, 4, 28, 38	30, 7, 14, 26	10, 26, 34, 31	10, 35, 17, 7	2, 6, 34, 10	35, 37, 10, 2	all	28, 15, 10, 36

(continued)

Appendix (CONTINUED) The Contradiction Matrix

Improved Feature	Worsened Feature									
	11	12	13	14	15	16	17	18	19	20
1 Weight of moving object	10, 36, 37, 40	10, 14, 35, 40	1, 35, 19, 39	28, 27, 18, 40	5, 34, 31, 35	all	6, 29, 4, 38	19, 1, 32	35, 12, 34, 31	all
2 Weight of stationary object	13, 29, 10, 18	13, 10, 29, 14	26, 39, 1, 40	28, 2, 10, 27	all	2, 27, 19, 6	28, 19, 32, 22	19, 32, 35	all	18, 19, 28, 1
3 Length of moving object	1, 8, 35	1, 8, 10, 29	1, 8, 15, 34	8, 35, 29, 34	19	all	10, 15, 19	32	8, 35, 24	all
4 Length of stationary object	1, 14, 35	13, 14, 15, 7	39, 37, 35	15, 14, 28, 26	all	1, 10, 35	3, 35, 38, 18	3, 25	all	all
5 Area of moving object	10, 15, 36, 28	5, 34, 29, 4	11, 2, 13, 39	3, 15, 40, 14	6, 3	all	2, 15, 16	15, 32, 19, 13	19, 32	all
6 Area of stationary object	10, 15, 36, 37	all	2, 38	40	all	2, 10, 19, 30	35, 39, 38	all	all	all
7 Volume of moving object	6, 35, 36, 37	1, 15, 29, 4	28, 10, 1, 39	9, 14, 15, 7	6, 35, 4	all	34, 39, 10, 18	2, 13, 10	35	all
8 Volume of stationary object	24, 35	7, 2, 35	34, 28, 35, 40	9, 14, 17, 15	all	35, 34, 38	35, 6, 4	all	all	all
9 Speed	6, 18, 38, 40	35, 15, 18, 34	28, 33, 1, 18	8, 3, 26, 14	3, 19, 35, 5	all	28, 30, 36, 2	10, 13, 19	8, 15, 35, 38	all
10 Force	18, 21, 11	10, 35, 40, 34	35, 10, 21	35, 10, 14, 27	19, 2	all	35, 10, 21	all	19, 17, 10	1, 16, 36, 37
11 Stress or pressure	all	35, 4, 15, 10	35, 33, 2, 40	9, 18, 3, 40	19, 3, 27	all	35, 39, 19, 2	all	14, 24, 10, 37	all

12	Shape	34, 15, 10, 14	all	33, 1, 18, 4	30, 14, 10, 40	14, 26, 9, 25	all	22, 14, 19, 32	13, 15, 32	2, 6, 34, 14	all
13	Stability of the object's composition	2, 35, 40	22, 1, 18, 4	all	17, 9, 15	13, 27, 10, 35	39, 3, 35, 23	35, 1, 32	32, 3, 27, 16	13, 19	27, 4, 29, 18
14	Strength	10, 3, 18, 40	10, 30, 35, 40	13, 17, 35	v	27, 3, 26	all	30, 10, 40	35, 19	19, 35, 10	35
15	Duration of action by a moving object	19, 3, 27	14, 26, 28, 25	13, 3, 35	27, 3, 10	all	all	19, 35, 39	2, 19, 4, 35	28, 6, 35, 18	all
16	Duration of action by a stationary object	all	all	39, 3, 35, 23	all	all	all	19, 18, 36, 40	all	all	all
17	Temperature	35, 39, 19, 2	14, 22, 19, 32	1, 35, 32	10, 30, 22, 40	19, 13, 39	19, 18, 36, 40	all	32, 30, 21, 16	19, 15, 3, 17	all
18	Illumination intensity	all	32, 30	32, 3, 27	35, 19	2, 19, 6	all	32, 35, 19	all	32, 1, 19	32, 35, 1, 15
19	Use of energy by moving object	23, 14, 25	12, 2, 29	19, 13, 17, 24	5, 19, 9, 35	28, 35, 6, 18	all	19, 24, 3, 14	2, 15, 19	all	all

(continued)

Appendix (CONTINUED) The Contradiction Matrix

Improved Feature		Worsened Feature									
		11	12	13	14	15	16	17	18	19	20
20	Use of energy by stationary object	all	all	27, 4, 29, 18	35	all	all	all	19, 2, 35, 32	all	all
21	Power	22, 10, 35	29, 14, 2, 40	35, 32, 15, 31	26, 10, 28	19, 35, 10, 38	16	2, 14, 17, 25	16, 6, 19	16, 6, 19, 37	all
22	Loss of energy	all	all	14, 2, 39, 6	26	all	all	19, 38, 7	1, 13, 32, 15	all	all
23	Loss of substance	3, 36, 37, 10	29, 35, 3, 5	2, 14, 30, 40	35, 28, 31, 40	28, 27, 3, 18	27, 16, 18, 38	21, 36, 39, 31	1, 6, 13	35, 18, 24, 5	28, 27, 12, 31
24	Loss of information	all	all	all	all	10	10	all	19	all	all
25	Loss of time	37, 36, 4	4, 10, 34, 17	35, 3, 22, 5	29, 3, 28, 18	20, 10, 28, 18	28, 20, 10, 16	35, 29, 21, 18	1, 19, 26, 17	35, 38, 19, 18	1
26	Quantity of substance/the matter	10, 36, 14, 3	35, 14	15, 2, 17, 40	14, 35, 34, 10	3, 35, 10, 40	3, 35, 31	3, 17, 39	all	34, 29, 16, 18	3, 35, 31
27	Reliability	10, 24, 35, 19	35, 1, 16, 11	all	11, 28	2, 35, 3, 25	34, 27, 6, 40	3, 35, 10	11, 32, 13	21, 11, 27, 19	36, 23
28	Measurement accuracy	6, 28, 32	6, 28, 32	32, 35, 13	28, 6, 32	28, 6, 32	10, 26, 24	6, 19, 28, 24	6, 1, 32	3, 6, 32	all
29	Manufacturing precision	3, 35	32, 30, 40	30, 18	3, 27	3, 27, 40	all	19, 26	3, 32	32, 2	all
30	External harm affects the object	22, 2, 37	22, 1, 3, 35	35, 24, 30, 18	18, 35, 37, 1	22, 15, 33, 28	17, 1, 40, 33	22, 33, 35, 2	1, 19, 32, 13	1, 24, 6, 27	10, 2, 22, 37

31	Object-generated harmful factors	2, 33, 27, 18	35, 1	35, 40, 27, 39	15, 35, 22, 2	15, 22, 33, 31	21, 39, 16, 22	22, 35, 2, 24	19, 24, 39, 32	2, 35, 6	19, 22, 18
32	Ease of manufacture	35, 19, 1, 37	1, 28, 13, 27	11, 13, 1	1, 3, 10, 32	27, 1, 4	35, 16	27, 26, 18	28, 24, 27, 1	28, 26, 27, 1	1, 4
33	Ease of operation	2, 32, 12	15, 34, 29, 28	32, 35, 30	32, 40, 3, 28	29, 3, 8, 25	1, 16, 25	26, 27, 13	13, 17, 1, 24	1, 13, 24	all
34	Ease of repair	13	1, 13, 2, 4	2, 35	11, 1, 2, 9	11, 29, 28, 27	1	4, 10	15, 1, 13	15, 1, 28, 16	all
35	Adaptability or versatility	35, 16	15, 37, 1, 8	35, 30, 14	35, 3, 32, 6	13, 1, 35	2, 16	27, 2, 3, 35	6, 22, 26, 1	19, 35, 29, 13	all
36	Device complexity	19, 1, 35	29, 13, 28, 15	2, 22, 17, 19	2, 13, 28	10, 4, 28, 15	all	2, 17, 13	24, 17, 13	27, 2, 29, 28	all
37	Difficulty of detecting and measuring	35, 36, 37, 32	27, 13, 1, 39	11, 22, 39, 30	27, 3, 15, 28	19, 29, 39, 25	25, 34, 6, 35	3, 27, 35, 16	2, 24, 26	35, 38	19, 35, 16
38	Extent of automation	13, 35	15, 32, 1, 13	18, 1	25, 13	6, 9	all	26, 2, 19	8, 32, 19	2, 32, 13	all
39	Productivity	10, 37, 14	14, 10, 34, 40	35, 3, 22, 39	29, 28, 10, 18	35, 10, 2, 18	20, 10, 16, 38	35, 21, 28, 10	26, 17, 19, 1	35, 10, 38, 19	1

(continued)

Appendix (CONTINUED) The Contradiction Matrix

Improved Feature		Worsened Feature									
		21	22	23	24	25	26	27	28	29	30
1	Weight of moving object	12, 36, 18, 31	6, 2, 34, 19	5, 35, 3, 31	10, 24, 35	10, 35, 20, 28	3, 26, 18, 31	1, 3, 11, 27	28, 27, 35, 26	28, 35, 26, 18	22, 21, 18, 27
2	Weight of stationary object	15, 19, 18, 22	18, 19, 28, 15	5, 8, 13, 30	10, 15, 35	10, 20, 35, 26	19, 6, 18, 26	10, 28, 8, 3	18, 26, 28	10, 1, 35, 17	2, 19, 22, 37
3	Length of moving object	1, 35	7, 2, 35, 39	4, 29, 23, 10	1, 24	15, 2, 29	29, 35	10, 14, 29, 40	28, 32, 4	10, 28, 29, 37	1, 15, 17, 24
4	Length of stationary object	12, 8	6, 28	10, 28, 24, 35	24, 26,	30, 29, 14	all	15, 29, 28	32, 28, 3	2, 32, 10	1, 18
5	Area of moving object	19, 10, 32, 18	15, 17, 30, 26	10, 35, 2, 39	30, 26	26, 4	29, 30, 6, 13	29, 9	26, 28, 32, 3	2, 32	22, 33, 28, 1
6	Area of stationary object	17, 32	17, 7, 30	10, 14, 18, 39	30, 16	10, 35, 4, 18	2, 18, 40, 4	32, 35, 40, 4	26, 28, 32, 3	2, 29, 18, 36	27, 2, 39, 35
7	Volume of moving object	35, 6, 13, 18	7, 15, 13, 16	36, 39, 34, 10	2, 22	2, 6, 34, 10	29, 30, 7	14, 1, 40, 11	25, 26, 28	25, 28, 2, 16	22, 21, 27, 35
8	Volume of stationary object	30, 6	all	10, 39, 35, 34	all	35, 16, 32, 18	35, 3	2, 35, 16	all	35, 10, 25	34, 39, 19, 27
9	Speed	19, 35, 38, 2	14, 20, 19, 35	10, 13, 28, 38	13, 26	all	10, 19, 29, 38	11, 35, 27, 28	28, 32, 1, 24	10, 28, 32, 25	1, 28, 35, 23
10	Force	19, 35, 18, 37	14, 15	8, 35, 40, 5	all	10, 37, 36	14, 29, 18, 36	3, 35, 13, 21	35, 10, 23, 24	28, 29, 37, 36	1, 35, 40, 18
11	Stress or pressure	10, 35, 14	2, 36, 25	10, 36, 3, 37	all	37, 36, 4	10, 14, 36	10, 13, 19, 35	6, 28, 25	3, 35	22, 2, 37

12	Shape	4, 6, 2	14	35, 29, 3, 5	all	14, 10, 34, 17	36, 22	10, 40, 16	28, 32, 1	32, 30, 40	22, 1, 2, 35
13	Stability of the object's composition	32, 35, 27, 31	14, 2, 39, 6	2, 14, 30, 40	all	35, 27	15, 32, 35	all	13	18	35, 24, 30, 18
14	Strength	10, 26, 35, 28	35	35, 28, 31, 40	all	29, 3, 28, 10	29, 10, 27	11, 3	3, 27, 16	3, 27	18, 35, 37, 1
15	Duration of action by a moving object	19, 10, 35, 38	all	28, 27, 3, 18	10	20, 10, 28, 18	3, 35, 10, 40	11, 2, 13	3	3, 27, 16, 40	22, 15, 33, 28
16	Duration of action by a stationary object	16	all	27, 16, 18, 38	10	28, 20, 10, 16	3, 35, 31	34, 27, 6, 40	10, 26, 24	all	17, 1, 40, 33
17	Temperature	2, 14, 17, 25	21, 17, 35, 38	21, 36, 29, 31	all	35, 28, 21, 18	3, 17, 30, 39	19, 35, 3, 10	32, 19, 24	24	22, 33, 35, 2
18	Illumination intensity	32	13, 16, 1, 6	13, 1	1, 6	19, 1, 26, 17	1, 19	all	11, 15, 32	3, 32	15, 19
19	Use of energy by moving object	6, 19, 37, 18	12, 22, 15, 24	35, 24, 18, 5	all	35, 38, 19, 18	34, 23, 16, 18	19, 21, 11, 27	3, 1, 32	all	1, 35, 6, 27

(continued)

Appendix (CONTINUED) The Contradiction Matrix

Improved Feature		21	22	23	24	25	26	27	28	29	30
						Worsened Feature					
20	Use of energy by stationary object	all	all	28, 27, 18, 31	all	all	3, 35, 31	10, 36, 23	all	all	10, 2, 22, 37
21	Power	all	10, 35, 38	28, 27, 18, 38	10, 19	35, 20, 10, 6	4, 34, 19	19, 24, 26, 31	32, 15, 2	32, 2	19, 22, 31, 2
22	Loss of Energy	3, 38	all	35, 27, 2, 37	19, 10	10, 18, 32, 7	7, 18, 25	11, 10, 35	32	all	21, 22, 35, 2
23	Loss of substance	28, 27, 18, 38	35, 27, 2, 31	all	all	15, 18, 35, 10	6, 3, 10, 24	10, 29, 39, 35	16, 34, 31, 28	35, 10, 24, 31	33, 22, 30, 40
24	Loss of Information	10, 19	19, 10	all	all	24, 26, 28, 32	24, 28, 35	10, 28, 23	all	all	22, 10, 1
25	Loss of Time	35, 20, 10, 6	10, 5, 18, 32	35, 18, 10, 39	24, 26, 28, 32	all	35, 38, 18, 16	10, 30, 4	24, 34, 28, 32	24, 26, 28, 18	35, 18, 34
26	Quantity of substance/matter	35	7, 18, 25	6, 3, 10, 24	24, 28, 35	35, 38, 18, 16	all	18, 3, 28, 40	13, 2, 28	33, 30	35, 33, 29, 31
27	Reliability	21, 11, 26, 31	10, 11, 35	10, 35, 29, 39	10, 28	10, 30, 4	21, 28, 40, 3	all	32, 3, 11, 23	11, 32, 1	27, 35, 2, 40
28	Measurement accuracy	3, 6, 32	26, 32, 27	10, 16, 31, 28	all	24, 34, 28, 32	2, 6, 32	5, 11, 1, 23	all	all	28, 24, 22, 26
29	Manufacturing precision	32, 2	13, 32, 2	35, 31, 10, 24	all	32, 26, 28, 18	32, 30	11, 32, 1	all	all	26, 28, 10, 36
30	External harm affects the object	19, 22, 31, 2	21, 22, 35, 2	33, 22, 19, 40	22, 10, 2	35, 18, 34	35, 33, 29, 31	27, 24, 2, 40	28, 33, 23, 26	26, 28, 10, 18	all

31	Object-generated harmful factors	2, 35, 18	21, 35, 2, 22	10, 1, 34	10, 21, 29	1, 22	3, 24, 39, 1	24, 2, 40, 39	3, 33, 26	4, 17, 34, 26	all
32	Ease of manufacture	27, 1, 12, 24	19, 35	15, 34, 33	32, 24, 18, 16	35, 28, 34, 4	35, 23, 1, 24		1, 35, 12, 18	all	24, 2
33	Ease of operation	35, 34, 2, 10	2, 19, 13	28, 32, 2, 24	4, 10, 27, 22	4, 28, 10, 34	12, 35	17, 27, 8, 40	25, 13, 2, 34	1, 32, 35, 23	2, 25, 28, 39
34	Ease of repair	15, 10, 32, 2	15, 1, 32, 19	2, 35, 34, 27	all	32, 1, 10, 25	2, 28, 10, 25	11, 10, 1, 16	10, 2, 13	25, 10	35, 10, 2, 16
35	Adaptability or versatility	19, 1, 29	18, 15, 1	15, 10, 2, 13	all	35, 28	3, 35, 15	35, 13, 8, 24	35, 5, 1, 10	all	35, 11, 32, 31
36	Device complexity	20, 19, 30, 34	10, 35, 13, 2	35, 10, 28, 29	all	6, 29	13, 3, 27, 10	13, 35, 1	2, 26, 10, 34	26, 24, 32	22, 19, 29, 40
37	Difficulty of detecting and measuring	18, 1, 16, 10	35, 3, 15, 19	1, 18, 10, 24	35, 33, 27, 22	18, 28, 32, 9	3, 27, 29, 18	27, 40, 28, 8	26, 24, 32, 28	all	22, 19, 29, 28
38	Extent of automation	28, 2, 27	23, 28	35, 10, 18, 5	35, 33	24, 28, 35, 30	35, 13	11, 27, 32	28, 26, 10, 34	28, 26, 18, 23	2, 33
39	Productivity	35, 20, 10	28, 10, 29, 35	28, 10, 35, 23	13, 15, 23	all	35, 38	1, 35, 10, 38	1, 10, 34, 28	18, 10, 32, 1	22, 35, 13, 24

(continued)

Appendix (CONTINUED) The Contradiction Matrix

Improved Feature		31	32	33	34	35	36	37	38	39
									Worsened Feature	
1	Weight of moving object	22, 35, 31, 39	27, 28, 1, 36	35, 3, 2, 24	2, 27, 28, 11	29, 5, 15, 8	26, 30, 36, 34	28, 29, 26, 32	26, 35, 18, 19	35, 3, 24, 37
2	Weight of stationary object	35, 22, 1, 39	28, 1, 9	6, 13, 1, 32	2, 27, 28, 11	19, 15, 29	1, 10, 26, 39	25, 28, 17, 15	2, 26, 35	1, 28, 15, 35
3	Length of moving object	17, 15	1, 29, 17	15, 29, 35, 4	1, 28, 10	14, 15, 1, 16	1, 19, 26, 24	35, 1, 26, 24	17, 24, 26, 16	14, 4, 28, 29
4	Length of stationary object	all	15, 17, 27	2, 25	3	1, 35	1, 26	26	all	30, 14, 7, 26
5	Area of moving object	17, 2, 18, 39	13, 1, 26, 24	15, 17, 13, 16	15, 13, 10, 1	15, 30	14, 1, 13	2, 36, 26, 18	14, 30, 28, 23	10, 26, 34, 2
6	Area of stationary object	22, 1, 40	40, 16	16, 4	16	15, 16	1, 18, 36	2, 35, 30, 18	23	10, 15, 17, 7
7	Volume of moving object	17, 2, 40, 1	29, 1, 40	15, 13, 30, 12	10	15, 29	26, 1	29, 26, 4	35, 34, 16, 24	10, 6, 2, 34
8	Volume of stationary object	30, 18, 35, 4	35	all	1	all	1, 31	2, 17, 26		35, 37, 10, 2
9	Speed	2, 24, 35, 21	35, 13, 8, 1	32, 28, 13, 12	34, 2, 28, 27	15, 10, 26	10, 28, 4, 34	3, 34, 27, 16	10, 18	all
10	Force	13, 3, 36, 24	15, 37, 18, 1	1, 28, 3, 25	15, 1, 11	15, 17, 18, 20	26, 35, 10, 18	36, 37, 10, 19	2, 35	3, 28, 35, 37
11	Stress or pressure	2, 33, 27, 18	1, 35, 16	11	2	35	19, 1, 35	2, 36, 37	35, 24	10, 14, 35, 37

12	Shape	35, 1	1, 32, 17, 28	32, 15, 26	2, 13, 1	1, 15, 29	16, 29, 1, 28	15, 13, 39	15, 1, 32	17, 26, 34, 10
13	Stability of the object's composition	35, 40, 27, 39	35, 19	32, 35, 30	2, 35, 10, 16	35, 30, 34, 2	2, 35, 22, 26	35, 22, 39, 23	1, 8, 35	23, 35, 40, 3
14	Strength	15, 35, 22, 2	11, 3, 10, 32	32, 40, 25, 2	27, 11, 3	15, 3, 32	2, 13, 25, 28	27, 3, 15, 40	15	29, 35, 10, 14
15	Duration of action by a moving object	21, 39, 16, 22	27, 1, 4	12, 27	29, 10, 27	1, 35, 13	10, 4, 29, 15	19, 29, 39, 35	6, 10	35, 17, 14, 19
16	Duration of action by a stationary object	22	35, 10	1	1	2	all	25, 34, 6, 35	1	20, 10, 16, 38
17	Temperature	22, 35, 2, 24	26, 27	26, 27	4, 10, 16	2, 18, 27	2, 17, 16	3, 27, 35, 31	26, 2, 19, 16	15, 28, 35
18	Illumination intensity	35, 19, 32, 39	19, 35, 28, 26	28, 26, 19	15, 17, 13, 16	15, 1, 19	6, 32, 13	32, 15	2, 26, 10	2, 25, 16
19	Use of energy by moving object	2, 35, 6	28, 26, 30	19, 35	1, 15, 17, 28	15, 17, 13, 16	2, 29, 27, 28	35, 38	32, 2	12, 28, 35

(continued)

Appendix (CONTINUED)   The Contradiction Matrix

Improved Feature		31	32	33	34	35	36	37	38	39	
20	Use of energy by stationary object	19, 22, 18	1, 4	all	all	all	all	19, 35, 16, 25	all	1, 6	
21	Power	2, 35, 18	26, 10, 34	26, 35, 10	35, 2, 10, 34	19, 17, 34	20, 19, 30, 34	19, 35, 16	28, 2, 17	28, 35, 34	
22	Loss of energy	21, 35, 2, 22	all	35, 32, 1	2, 19	all	7, 23	35, 3, 15, 23	2	28, 10, 29, 35	
23	Loss of substance	10, 1, 34, 29	15, 34, 33	32, 28, 2, 24	2, 35, 34, 27	15, 10, 2	35, 10, 28, 24	35, 18, 10, 13	35, 10, 18	28, 35, 10, 23	
24	Loss of information	10, 21, 22	32	27, 22	all	all	all	35, 33	35	13, 23, 15	
25	Loss of time	35, 22, 18, 39	35, 28, 34, 4	4, 28, 10, 34	32, 1, 10	35, 28	6, 29	18, 28, 32, 10	24, 28, 35, 30	all	
26	Quantity of substance/ matter	3, 35, 40, 39	29, 1, 35, 27	35, 29, 25, 10	2, 32, 10, 25	15, 3, 29	3, 13, 27, 10	3, 27, 29, 18	8, 35	13, 29, 3, 27	
27	Reliability	35, 2, 40, 26	all	27, 17, 40	1, 11	13, 35, 8, 24	13, 35, 1	27, 40, 28	11, 13, 27	1, 35, 29, 38	
28	Measurement accuracy	3, 33, 39, 10	6, 35, 25, 18	1, 13, 17, 34	1, 32, 13, 11	13, 35, 2	27, 35, 10, 34	26, 24, 32, 28	28, 2, 10, 34	10, 34, 28, 32	
29	Manufacturing precision	4, 17, 34, 26	all	1, 32, 35, 23	25, 10	all	26, 2, 18	all	26, 28, 18, 23	10, 18, 32, 39	
30	External harm affects the object	all	24, 35, 2	2, 25, 28, 39	35, 10, 2	35, 11, 22, 31	22, 19, 29, 40	22, 19, 29, 40	33, 3, 34	22, 35, 13, 24	

	31	32	33	34	35	36	37	38	39
31 Object-generated harmful factors	all	all	all	all	all	19, 1, 31	2, 21, 27, 1	2	22, 35, 18, 39
32 Ease of manufacture	all	all	2, 5, 13, 16	35, 1, 11, 9	2, 13, 15	27, 26, 1	6, 28, 11, 1	8, 28, 1	35, 1, 10, 28
33 Ease of operation	all	2, 5, 12	all	12, 26, 1, 32	15, 34, 1, 16	32, 26, 12, 17	all	1, 34, 12, 3	15, 1, 28
34 Ease of repair	all	1, 35, 11, 10	1, 12, 26, 15	all	7, 1, 4, 16	35, 1, 13, 11	all	34, 35, 7, 13	1, 32, 10
35 Adaptability or versatility	all	1, 13, 31	15, 34, 1, 16	1, 16, 7, 4	all	15, 29, 37, 28	all	27, 34, 35	35, 28, 6, 37
36 Device complexity	19, 1	27, 26, 1, 13	27, 9, 26, 24	1, 13	29, 15, 28, 37	all	15, 10, 37, 28	15, 1, 24	12, 17, 28
37 Difficulty of detecting and measuring	2, 21	5, 28, 11, 29	2, 5	12, 26	1, 15	15, 10, 37, 28	all	34, 21	35, 18
38 Extent of automation	2	1, 26, 13	1, 12, 34, 3	1, 35, 13	27, 4, 1, 35	15, 24, 10	34, 27, 25	all	5, 12, 35, 26
39 Productivity	35, 22, 18, 39	35, 28, 2, 24	1, 28, 7, 10	1, 32, 10, 25	1, 35, 28, 37	12, 17, 28, 24	35, 18, 27, 2	5, 12, 35, 26	all

Modified from Altshuller, G.S., *40 Principles*, TIC, Worcester, 1997, 135. With permission.

11

EVALUATION OF THE MODEL
FOR PROBLEM SOLVING

In the Preface, we stressed that the reader should test and refine the generic model for problem solving. Accept TRIZ because it works, not because it is in fashion.

When you use TRIZ, you will have two ways to evaluate the process:

1. Evaluate the results of implementation against the criteria of the *ideal final result*, as studied in Chapter 7.
2. Evaluate the model and tools of TRIZ against your own accumulated knowledge and experience.

Using the model for problem solving provides a structured way to arrange and organize the TRIZ tools, and also gives you a system for creating new knowledge. Contradiction, ideality and other basic concepts are open for further analysis, deepening and implementation in new areas. The model brings a component of research to your work. TRIZ is evolving continuously; it is not a frozen dogma that does not change and is used without question. One should ask repeatedly: "Why this model? Why this tool? Why not something else?"

Each time that you use TRIZ to solve a problem, you have the opportunity to improve your own learning and to help improve TRIZ for others. Some people add yet another question to the agenda in Chapter 8: "What did I do well in this application of TRIZ? What should I do differently next time?"

We hope that you have implemented the models and tools in this book to create innovative solutions to your problems. When you have done that, you can use the exercise in Table 11.1 to contribute to the development of TRIZ.

Table 11.1   Exercise: Evaluation of the Model for Problem Solving.

1. Which concepts, models and tools are most valid and useful for my work?

_____

_____

_____

2. Which points may require further development?

_____

_____

_____

3. Other thoughts

_____

_____

_____

You can share these thoughts within your own organization, and improve the version of TRIZ that you are using, or you can share them with the global TRIZ community.[1]

REFERENCE

1. You are encouraged to e-mail your suggestions to the authors at: http://www.triz-journal.com/SimplifiedTRIZ/

12

HOW TO IMPROVE BUSINESS WITH TRIZ

If you are a manager or a consultant and have heard of TRIZ, you probably have a question: "How can I prove that TRIZ really works and will improve our business?"

If you have already worked your way through the first 11 chapters and tried the exercises, you are already proficient at using TRIZ by yourself. This generates a new problem: how to help other people in the organization adopt TRIZ.

Those of you who work in organizations, whether they be private companies, government agencies or mom-and-pop operations, should take TRIZ back to your co-workers to get the benefit of solving your own problems in your own environment, where you can develop these new ideas. Private companies will be looking for a competitive advantage — to patent, to set new standards, or to get proprietary advantage from being first to market with new concepts — and public organizations will be seeking maximum benefit for minimum cost. Both kinds of organizations need the breakthrough creativity of TRIZ.

This chapter and those that follow will help you in all these cases. The whole book can be divided roughly into three parts:

1. How to solve problems by analyzing contradictions (Chapters 1–7)
2. The improvement of systems using the patterns of evolution and 40 principles (Chapters 9–10)
3. The business applications of TRIZ (Chapters 12–14), where the systematic implementation process for TRIZ is presented to make the benefits of TRIZ quickly available to your organization.

In the last section, we discuss how to integrate TRIZ with other tools and methodologies. The following two chapters (13–14) demonstrate in detail the use of TRIZ with two important systems of improvement, the theory of constraints (TOC) and Six Sigma.

In this chapter, we first consider the most typical obstacles to adoption of TRIZ. Second, we tell how to overcome the obstacles and present the flowchart for the introduction of TRIZ. In the third section, we study three main steps of the introduction.

12.1 TYPICAL OBSTACLES TO THE ADOPTION OF TRIZ

The same obstacles to the implementation of TRIZ are met in large, small, public and private organizations. The ways to overcome the obstacles are very similar, too, which makes knowledge of the typical obstacles very useful.

TRIZ will give your organization the capability for breakthrough solutions to difficult problems. TRIZ radically enhances the quality and quantity of idea generation. Reading this statement, people have one of two reactions:

1. If it sounds too good to be true, it probably is. Just one more piece of hype. Management training "flavor of the month."
2. Great! Let's get it in here and start everybody using it immediately.

If you have used TRIZ to generate creative solutions to your own problems, you will avoid reaction number 1. But what can you offer to people who have not experienced the power of TRIZ?

The primary obstacles to organization-wide adoption of TRIZ are human, not technical:

■ *Time*: People are too busy fighting fires to learn new methods of fire prevention.
■ *Suspicion*: Other "new methods" have over-promised productivity improvement, customer satisfaction, faster time to market, higher return on investment (ROI) or economic value added (EVA), etc.
■ *Traditional systems of project management:* If traditional milestones measure the success of a project and the new process doesn't match those milestones, there will be great pressure to work within the existing system. For example, some product development systems have a time period dedicated to conducting tradeoff studies. TRIZ tells us that good solutions to problems avoid tradeoffs. If the organization continues to mandate the use of its traditional system, the use of TRIZ will be discouraged.

■ *The NIH syndrome.* NIH means *"not invented here"* and can have double meaning for TRIZ — both "not invented in the organization" and "not invented in this country."

■ *"Well, it may work for so-and-so, but it won't work for us."* "Our problems are different/high-tech/not in their database/controlled by regulators, etc." This comment is actually a subcategory of NIH.

12.2 HOW TO INTRODUCE TRIZ INTO YOUR ORGANIZATION

The flowchart in Figure 12.1 describes a structured method of introducing TRIZ into organizations that overcomes these obstacles. This method uses no "tricks" of cultural change or subtleties of organizational dynamics. It gets the professionals and managers in service, engineering, production and distribution to experience TRIZ immediately and helps them get breakthrough results on their own problems quickly. Of course, some organizations are far less structured than this — we've even seen success in places where the company just buys everyone a copy of this book.

The effect of these immediate increases in creativity is that the obstacles labeled "suspicion" and "NIH" are removed and the organization then uses its own resources (and its enhanced creativity) to reallocate the time of key people. In the early stages of TRIZ implementation, having a lot of new ideas is not always seen as a benefit, because the organization may lack the resources to follow up on all of them.[1] The rationale for the process represented by the flowchart will be clear if each step is analyzed in terms of its direct results — new ideas, new concepts, creativity improvements — and its organizational change results.

12.3 IMPLEMENTING THE STEPS OF THE FLOWCHART

The flowchart is a detailed outline of the steps that lead to successful TRIZ implementation. Three major steps encompass all the details:

1. The decision is made that increased innovation is needed.
2. Pilot projects. TRIZ and methods of teaching TRIZ are tested on the organization's projects, with its people working in their own surroundings. Implementation of the results of the pilot projects is a key to success, because it will let the entire organization see how the TRIZ solutions work in practice.
3. Acceptance. TRIZ becomes part of the normal methods of operation for the organization.

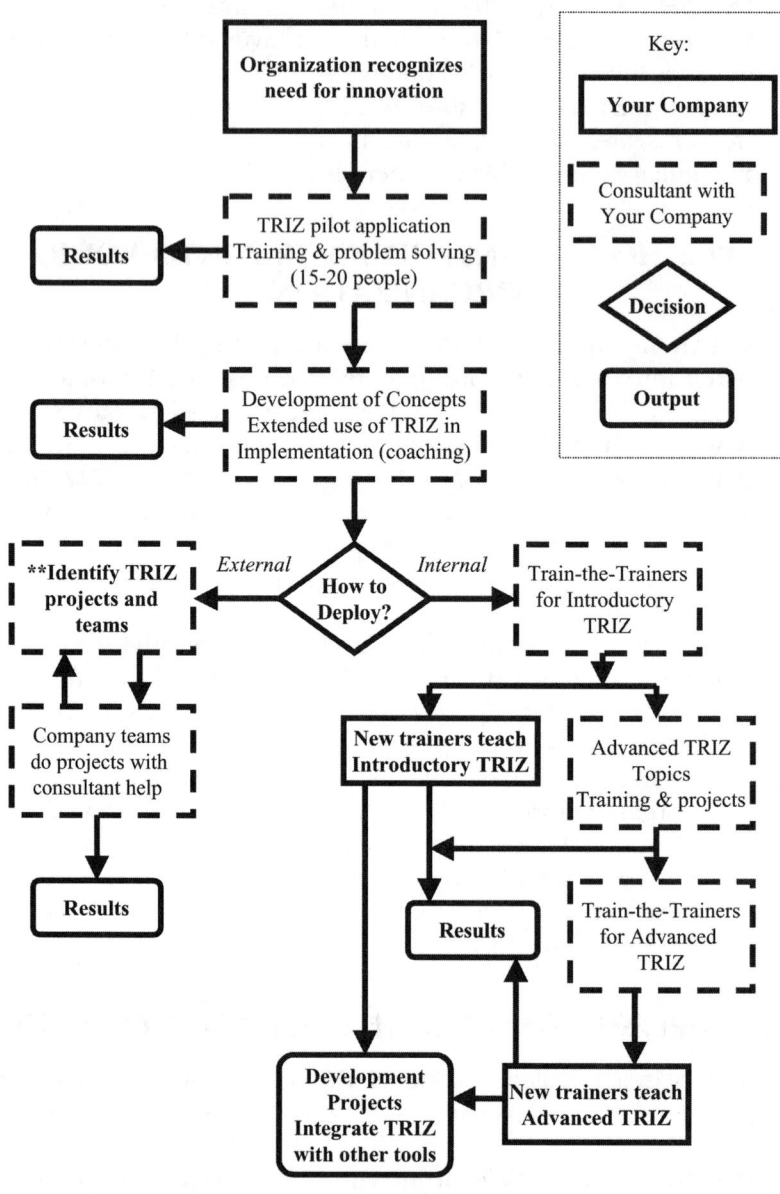

Figure 12.1 Flow chart for TRIZ implementation. This, the most structured level, is for large companies that want to become completely self-sufficient. Companies that want to use TRIZ quickly without studying it will start at the box marked **. The *internal* path is for the development of TRIZ experts inside the organization. The *external* path uses outside experts. Choosing whether to use consultants and outside experts is discussed in the text.

12.3.1 Step 1

Step 1 is the organization's decision that increased innovation is needed. Most commonly, in private business, this comes because of competitive pressure or, in public agencies, from citizen demand, although occasionally it comes from regulatory requirements. One or more organization managers is selected to be the TRIZ "champion" to orchestrate the introduction and institutionalization of TRIZ. Frequently, these champions are people who have learned TRIZ and understand how it will help the organization and have used their knowledge of the organization's politics to get themselves appointed to do the job. It is more important that the champion be a respected person who understands how to get things done within the organization, than to have a person who knows TRIZ.

It is essential for the champion to clarify needs and deal with the obstacles. One way to start is by asking and answering these questions:

- **Do we really need increased innovation?** Champions may have to answer this question first, or get the leadership of the company to answer it, before dealing with TRIZ as the systematic way to increase innovation. The motivation for increased innovation could come from customers, from competitors or from a regulatory situation. Champions will use the answer to this question to help people understand why a new method is being introduced into the organization.
- **Will TRIZ work in our circumstances?** Companies that have been the first to implement generic models and templates of TRIZ in their industries (electronics, chemical and medical industry, food industry, financial services, insurance and others) have gotten great results. Because the TRIZ examples in existing textbooks did not deal with their particular industry, they had to work harder to understand how to apply it to their situation. Now that TRIZ has been applied in so many different circumstances, this problem is somewhat lessened, but the champions may still have to find examples to convince others that this new method can work.

  Commitment generates results and results strengthen commitment. The system presented here is designed to get results fast, so the commitment can grow. The best tools won't give good results if there is no clear understanding of the need to change from the previous way of doing things to a new system. This list of questions may help clarify the need to change:
- **What are the actual problems in our organization?** (For example: difficulties anticipating customers' needs)
- **What are our strengths?** (For example: good knowledge in science and engineering, strong financial analysis skills, excellent distribution

management, reputation for a caring attitude as well as strong medical skills, etc.)

■ **What are our weaknesses?** (For example: high-level experts have trouble finding simple solutions that customers understand, services and products are introduced with failure modes that cause customer problems, not all employees understand customer service is a priority, etc.)

■ **What is needed to enhance strengths and remove weaknesses?** (For example: providing experts with models that help them use their knowledge more effectively.)

See also Chapter 1: Why do people seek new ways to solve problems?

12.3.2   Step 2

Step 2 is the selection of one or more pilot projects for TRIZ introduction. Candidate pilot projects can come from competitive situations, regulatory changes, or from the organization's problem identification and corrective-action system. The "champion" selects those problems that will have the best combination of high-value payoff and usefulness as future teaching cases. Good pilot projects are those that are regarded as hard problems worth solving. These projects or problems are used as the case studies for an introductory class. Sometimes, the champion asks each class member to select a problem. The following memo was used at one company to help people select good projects to bring to their first TRIZ class:

TRIZ Project Selection

> TRIZ has many techniques for finding innovative solutions to hard problems in product, process and transaction situations. Here is the list of characteristics of "good" problems for the TRIZ classes:

> We know who the customer is, what the customer's needs are and why the present system is not satisfying those needs.

> We understand the root cause of the problem, not just the symptoms. BUT we don't know what to do.

> Why don't we know what to do?

> Sometimes, it is because of the presence of contradictions in the system.

> Sometimes, it is because the system has reached the limits of what can be achieved with the current technology or methodology — strength of materials, bandwidth, communications, or ....

Sometimes, we don't know where to start.

In your TRIZ class next week, you will not only learn the TRIZ methods, you will apply those methods to solve problems. Please bring anything you need to explain the problem — documents, drawings, etc. To give you some ideas about the kinds of problems that you should consider, here is a list of problems that have been solved by people in previous TRIZ classes:

Transaction: The project managers complain that the project management software takes too much time to use, so they don't keep the data current. Then, they lose the ability to do dynamic scheduling because their data are not current. Find a way to keep all the data current without taking extra time.

Transaction: The old telephone system for a large consulting company kept track of each call so that the appropriate client could be billed for the expenses. The new system doesn't have that capability, but some of the contracts require separate billing of phone calls. (Should have thought of this before installing the new system.) What can be done?

Business: The company has a standard method for deciding which new projects should be funded. Many people think it is too complicated and they find ways to bypass it, causing great confusion about which projects are funded and about how the decisions are made. What can be done?

Measurement: The process has been through several cycles of improvement and the yield has increased by several orders of magnitude. Measuring defects is required by the customer and it now takes a very large sample of the material to find enough defects to measure. Find a way to get a measurement that will be accepted by the customer without sacrificing a large quantity of product.

Measurement: The customer specification requires that measurements be made during the cooling of the product. But, inserting the thermometer causes damage to the part of the product where the thermometer is inserted. Find a way to comply with the customer specification without wasting product.

Process: A machine was originally designed to handle sheets of metal separated by sheets of paper at very high speed. It is now being used for sheets of metal without the paper and it is causing unacceptable cosmetic damage to the sheets. Reintroducing the paper is not possible, because of other processes downstream. Find a way to handle the sheets at high speed.

Process: A system produces a chemical product and a stream of waste material mixed with water. The waste has to be removed before the water can be recycled or disposed of. The present system requires three purification systems (one active, one being cleaned, one on standby) and an expensive, time-consuming method of cleaning (shovel the purifying material into a truck, transport it to a reprocessing facility, shovel it back into the truck, get it out of the truck and back into the system). Reduce the cost of the product by finding a way to make the waste purification system less expensive.

Product: A food wrapper must prevent grease from penetrating, both for sanitary reasons and to look good. But, ink (which is very much like grease) must stick to the wrapper, so that the product can be identified and advertised. How should the wrapper be constructed?

The flowchart shows several boxes with dashed lines, noted as "consultant, with your company." The champion will need to decide whether to use a consultant and, if the decision is "yes," will then need to select one. See Table 12.1 for a summary of the reasons to work with or without a consultant.

If you choose not to use a consultant, you can still use the flowchart shown in Figure 12.1. You will, however, need to get TRIZ training by other means, such as attending public seminars and conferences, reading TRIZ books and research papers, etc.[2]

If you decide to use a consultant, there are many resources available for finding one. Consultants and trainers populate the annual meetings of the Altshuller Institute and the European TRIZ Association. (See the Calendar page of *The TRIZ Journal* for dates and programs of the meetings.) Many of the authors of the articles in *The TRIZ Journal* are active consultants who include contact information in their articles.

Once you have decided on pilot projects and whether to use an outside consultant, you are ready to conduct the pilot project TRIZ class. We recommend an experiential style of teaching, in which the instructor teaches the basic principles, then coaches the class participants to solve real problems. This style has multiple benefits:

- Concepts for inventive solutions are generated for the selected projects or problems.
- The participants themselves generate the results.
- The participants learn to sort the results and get immediate and long-term benefits.

Table 12.1   Should You Use a Consultant to Help Introduce TRIZ to Your Organization?

Yes	No
If you select a TRIZ training expert, it saves more time than it would take for your staff to learn TRIZ and develop training materials and methods.	Costs more than having a small number of employees study on their own.
Improves confidence of the pilot project participants because the consultant can show successful results from other organizations.	Some organizations reject anything that comes from outside, based on bad past experiences with consultants.
Produces more sophisticated results based on the consultant's experience. The beginners become advanced practitioners of TRIZ much more quickly than if they learn only from their own study and projects.	It may take time to educate the consultant about the company's culture and problems.

■ The participants learn the TRIZ methodology well enough to apply it themselves.

Implementation of the results of the pilot project is very important. Because many new problems are generated during implementation, the class participants get to use their new skills and receive valuable reinforcement of what they have learned. The TRIZ results are visible to the organization, so that resistance to introduction of the new methods is reduced or eliminated.

After the successes in Step 2, the TRIZ "champion" and the organization's leadership pick one of two paths. The *internal* path produces a full team of internal TRIZ practitioners who replace the external instructors and consultants as their skills increase. The *external* path uses consultants to coach each team as each project is identified. Hybrid approaches, in which the external path is followed for quick results, have also been used successfully, while the internal path is followed for development of future self-sufficiency. The external path is also frequently used for strategic planning, for applications of TRIZ to technology forecasting for the entire industry and for product platforms during the time that it takes to develop the internal path. The internal experts learn the strategic uses of TRIZ and, during their advanced topics education, become internal consultants as well as instructors.

12.4 GAINING ADDITIONAL BENEFITS BY INTEGRATING TRIZ WITH OTHER METHODOLOGIES

12.4.1 Step 3

Step 3 is acceptance. As the organization develops its own internal experts, they take the lead in the integration of TRIZ with the organization's other methodologies. They become the collective "champions" in overcoming the last obstacle to TRIZ implementation: the traditional systems of project management. TRIZ will impact new product projects, process improvement projects and process re-engineering projects. Following the right-hand internal branch of the flowchart will overcome this obstacle. As more and more people learn TRIZ and as the organization develops its internal cadre of experts, they will integrate TRIZ with all the company's other tools. Other examples of the integration of TRIZ with existing tools include the following:

■ *TRIZ/QFD*. Quality function deployment (QFD) identifies and prioritizes the *voice of the customer* and the capabilities of the organization's technologies, then helps prioritize new concepts for design and production of products and services. TRIZ helps create the new concepts and resolve contradictions.[3–5]

■ *TRIZ/Robust design/Taguchi methods*. Robust design finds the right parameters to minimize all forms of waste and cost. TRIZ finds ways of creating the processes that will achieve those parameters.[6]

■ *TRIZ/DFM-A*: Design for manufacturability and design for assembly identify and prioritize features of design that make manufacturing and assembly low-cost, high-yield and short-cycle time. TRIZ resolves the technical problems encountered when implementing these features. Similarly, many organizations have developed their own guidelines for "design for serviceability," which is enhanced by TRIZ creativity in achieving serviceable designs.

■ *TRIZ/Concurrent Engineering* (or integrated product and process engineering, or product development teams, or supplier/developer/customer teams) These project management teams will use TRIZ at many levels ranging from technology forecasting to conceptual design to production design, from implementation problem solving to service, delivery and repair improvements.[7,8]

The relationship between TRIZ and QFD is best illustrated by the QFD matrix called the "house of quality."[9] After the QFD team has collected information by interviewing and observing the customer, the data are organized in a matrix, shown in Figure 12.2. Figure 12.3 indicates regions of the QFD house of quality matrix that signal the need for TRIZ. There

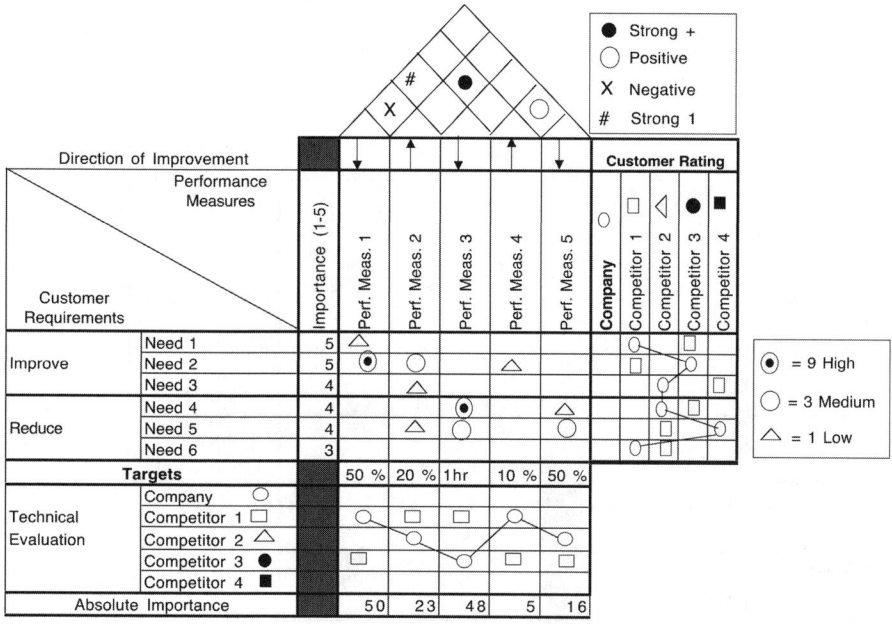

Figure 12.2 A typical QFD house of quality matrix. See Reference 9 to learn QFD methods. This matrix shows a strong correlation between Need 2 and Performance Measure 1, a medium correlation between Need 2 and Performance Measure 2 and a weak correlation between Need 2 and Performance Measure 4. There is strong conflict between Performance Measures 1 and 3 and positive reinforcement between Performance Measures 2 and 4.

are five obvious opportunities, marked on the matrix with an X, for interaction between QFD and TRIZ.

- ▪ Box 8 — Resolve conflict between performance measures.
- ▪ Box 5 — Empty rows. Use TRIZ to develop a means of satisfying customer needs.
- ▪ Box 5 — Empty columns. Use TRIZ to eliminate unnecessary activities. (Caution: some actions may be necessary for regulatory reasons not obvious to the end user.)
- ▪ Box 4 — Use TRIZ to develop performance measures and measurement methods (also applies to box 7)

For example, when designing a house, customers might say they want it to feel spacious but to take very little time to clean. These would be "customer needs" in box 1. The "performance measure" in box 4 that corresponds to spaciousness might be the volume or the area of the

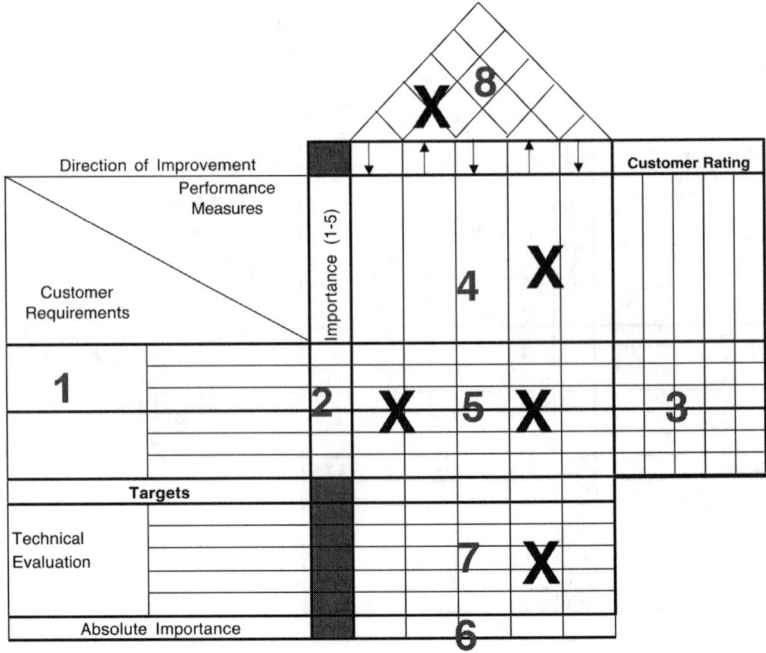

Figure 12.3 ✕ indicates parts of the QFD house of quality matrix that signal the need for TRIZ.

rooms. This would be in conflict with making the house very fast to clean, if that is done by making the rooms small. This would result in a contradiction notation in box 8. TRIZ would then be used to resolve the contradiction. It could be treated as an inherent contradiction:

- The room should be large (to feel spacious) but it should be small (to be quick to clean) or it could be treated as tradeoff contradiction.
- When the room gets bigger, it takes longer to clean.

Similarly, the other QFD matrices, such as the cost deployment matrix, the production (or service) planning matrix, the reliability matrix, etc., each have areas that will indicate, to those who are experienced in the use of QFD, the need for TRIZ.

The next two chapters will show extended examples of the use of TRIZ with the Theory of Constraints (TOC) and TRIZ with Six Sigma and other quality improvement methods.

At this level of integration, TRIZ passes from being seen as a tool, or a system of tools and methods and becomes an intrinsic part of an organization's method of gaining competitive advantage and fulfilling customer needs. Until it reaches this point, it will require nurturing and "championship" to keep people aware of their opportunities to apply TRIZ.

12.5 SUMMARY

- The major steps for bringing TRIZ into an organization are recognition of the need for increased innovation, using TRIZ for pilot projects and acceptance of TRIZ.

Table 12.2   A Worksheet for TRIZ Implementation

Answer these questions to begin planning your TRIZ implementation.

1.	In my organization, who would be a good champion?
2.	Will that person need a higher level management sponsor?
3.	If "yes," who would be a good sponsor?
4.	What will be the obstacles in my organization?
5.	What are the organization's strengths that TRIZ will increase?
6.	What are the organization's weaknesses that TRIZ will help overcome?
7.	Will we need to gather information about other organizations' successes with TRIZ to convince people that it can work in our company? If so, who will do the work of getting this information?[2]

- The organization's leadership will need to decide, depending on the organization's culture and on the time available, whether to use consultants or have their own employees do the entire process, or whether to use a hybrid of consultants and employees.
- Integration of TRIZ with other methods already in use in the organization, such as QFD, project management, design for manufacturability, etc., will accelerate its acceptance.

Use the worksheet in Table 12.2 to begin planning your TRIZ implementation.

REFERENCES

1. Cowley, M. and Domb, E., *Beyond Strategic Vision: Effective Corporate Action with Hoshin Planning*, Butterworth-Heinemann, Boston 1997, chap. 2.
2. In the U.S., the annual conference of the Altshuller Institute has tutorial sessions as well as research sessions. www.aitriz.org. The European TRIZ Association holds an annual meeting that is primarily research oriented, www.etria.net. Both meetings are good opportunities to learn TRIZ and to meet consultants and people from other companies to share experiences. See the Calendar section of *The TRIZ Journal*, www.triz-journal.com, for current dates and listings of public courses and meetings of other professional societies that have TRIZ programs.
3. Schlueter, M., QFD by TRIZ, *The TRIZ Journal*, June, 2001, www.triz-journal.com.
4. Domb, E. and Corbin, D., QFD, TRIZ and Entrepreneurial Intuition The DelCor Interactives International Case Study, *The TRIZ Journal*, September, 1998, www.triz-journal.com.
5. León-Rovira, N. and Aguayo, H., A new Model of the Conceptual Design Process using QFD/FA/TRIZ, *The TRIZ Journal*, July, 1998, www.triz-journal.com.
6. Matthew Hu, M., Yang, K. and Taguchi, S., Enhancing Robust Design with the Aid of TRIZ and Axiomatic Design, *The TRIZ Journal,* October and November, 2000, www.triz-journal.com.
7. Cavallucci, D. and Lutz, P., Intuitive Design Method (IDM), A New Approach on Design Methods Integration, *The TRIZ Journal*, October, 2000, www.triz-journal.com.
8. Zeidner, L. and Wood, R., The Collaborative Innovation (CI) Process, *The TRIZ Journal*, June, 2000, www.triz-journal.com.
9. Terninko, J., *Step by Step QFD*. Responsible Management, Nottingham,1995.

13

USING TRIZ WITH THE
THEORY OF CONSTRAINTS

Eliyahu Goldratt introduced an integrated problem-solving tool set loosely known as the "Theory of Constraints" (TOC) in the early 1990s. For those people who are already familiar with the TOC methods, this chapter is intended to show how TRIZ and TOC integrate very naturally. For those who would like more information on TOC, we suggest Goldratt's books, particularly,[1-3] H. William Dettmer's books[4,5] and the articles in *The TRIZ Journal* by Domb and Dettmer[6] and Ed Moura.[7]

The conflict resolution diagram (CRD), or "evaporating cloud," is one of the most powerful tools in the TOC tool set for resolving conflict. It is one of the few methods designed for formally structuring "win–win" solutions. In that respect, it is similar to TRIZ, in that both reject tradeoffs.

The strength of the CRD lies in two characteristics. First, it is an excellent way to structure and illustrate graphically the crucial elements of any conflict, starting from the overt indications and tracing the roots of the conflict back through the ultimate objectives of each side. Second, it helps to expose and identify the unspoken assumptions underlying each element of the conflict. Knowing what these assumptions are is the key to resolving the conflict in a "win–win" manner.

But, like most tools, the CRD is not perfect. While it is strong in the areas mentioned above, it is also somewhat weak in one key area: idea generation. The whole purpose of the CRD is to get at an idea for resolving the conflict, which is called an "injection" in the vocabulary of TOC. But this is the one aspect of using the CRD that could use some help. For generating injections, Goldratt has offered the idea of a reference environment (also called an alternative environment). While this approach can be effective on some kinds of problems, much as brainstorming, it leaves something to be desired for many people. Complementing the CRD, TRIZ offers a structured approach to the generation of ideas. Given the remarkable "fit" between the two tools, it seems obvious to combine the two techniques.

The CRD is composed of five elements: a common objective, two nonconflicting requirements and two conflicting prerequisites (See Figure 13.1). The requirements are necessary actions that must take place for the objective to be achieved. Each prerequisite is necessary for one of the requirements to be fulfilled. The essence of achieving "win–win" solutions lies in the idea that both requirements are satisfied, not necessarily both prerequisites.

To use the CRD to fashion a "win–win" solution, normally the conflicting prerequisites are articulated, then the requirements they support

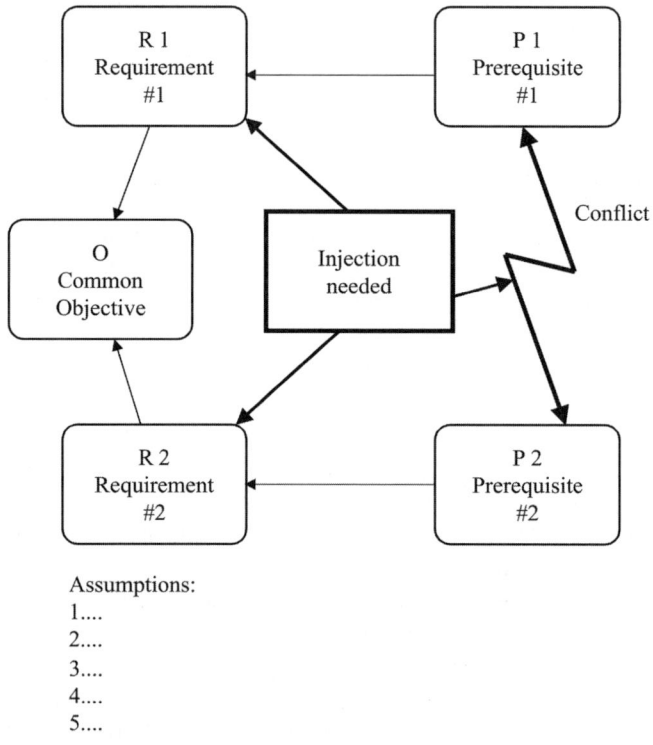

Figure 13.1   The Conflict Resolution Diagram. R1 and R2 are requirements for meeting the objective O. P1 is the prerequisite for R1 and P2 is the prerequisite for R2, but there is a conflict between P1 and P2. An idea, called an "injection" is needed to resolve the conflict. Modified from Dettmer, H.W.[6] With permission.

and the common objective of the two requirements are expressed. After these five elements are in place, the assumptions associated with each leg of the diagram are "coaxed out" into the open. The objective of this effort is to identify the assumptions that are either faulty to begin with or that might be rendered invalid by some other alternative action. Finally, when all the assumptions are exposed for each leg of the diagram and the vulnerable ones identified, an injection (idea for a solution) is created, usually to replace one or both of the conflicting prerequisites. It is clear that the CRD helps find contradictions and conflicts and helps clarify the reasons for the conflict, while TRIZ problem solving removes the conflict.

To demonstrate how the CRD and TRIZ might function effectively together, let's look at a complex example: the 1986 tragedy in the American space program — the Challenger accident. (This example is taken from an article by Ellen Domb and Bill Dettmer from the May 1999 issue of The TRIZ Journal.)

13.1   THE CHALLENGER CURRENT REALITY TREE

Almost everybody knows something about the causes behind the Challenger accident, but most people don't realize that the critical root cause was not the infamous "O-rings" that received such attention from the press. The real cause was much deeper than that. The chain of cause and effect that culminated in the explosion of the Challenger on January 28, 1986 began in 1972 with NASA's acquisition policies. Figure 13.2a is a representation of the factual situation in the form of a "current reality tree." For the purposes of our example, only the lower levels of the tree are shown here.

The current reality tree is a TOC tool that can be used to describe a complex situation as a chain of conflict resolution diagrams. Look for arrows that pass through an oval to another box. Read this as "IF the first box, AND the second box, … all happen, then the box at the end of the arrow will happen." The critical root cause is then the box that, if it is removed or modified, will prevent the result from happening. The boxes are labeled starting at 100 to make it easy to insert new causes when they are found as the research into the problem progresses.

Like most complex problem situations, especially vehicle accidents, many factors contributed to the Challenger disaster and the deadly chain of cause and effect might have been interrupted at several key points. One of these points dated back to 1977. The contractor selected to provide the solid rocket boosters (SRBs) for the Space Shuttle had been awarded the contract based primarily on the low cost of its bid. The contractor was able to submit such a low bid because its design concept involved scaling up in size the design for the Titan III solid rocket booster, a proven,

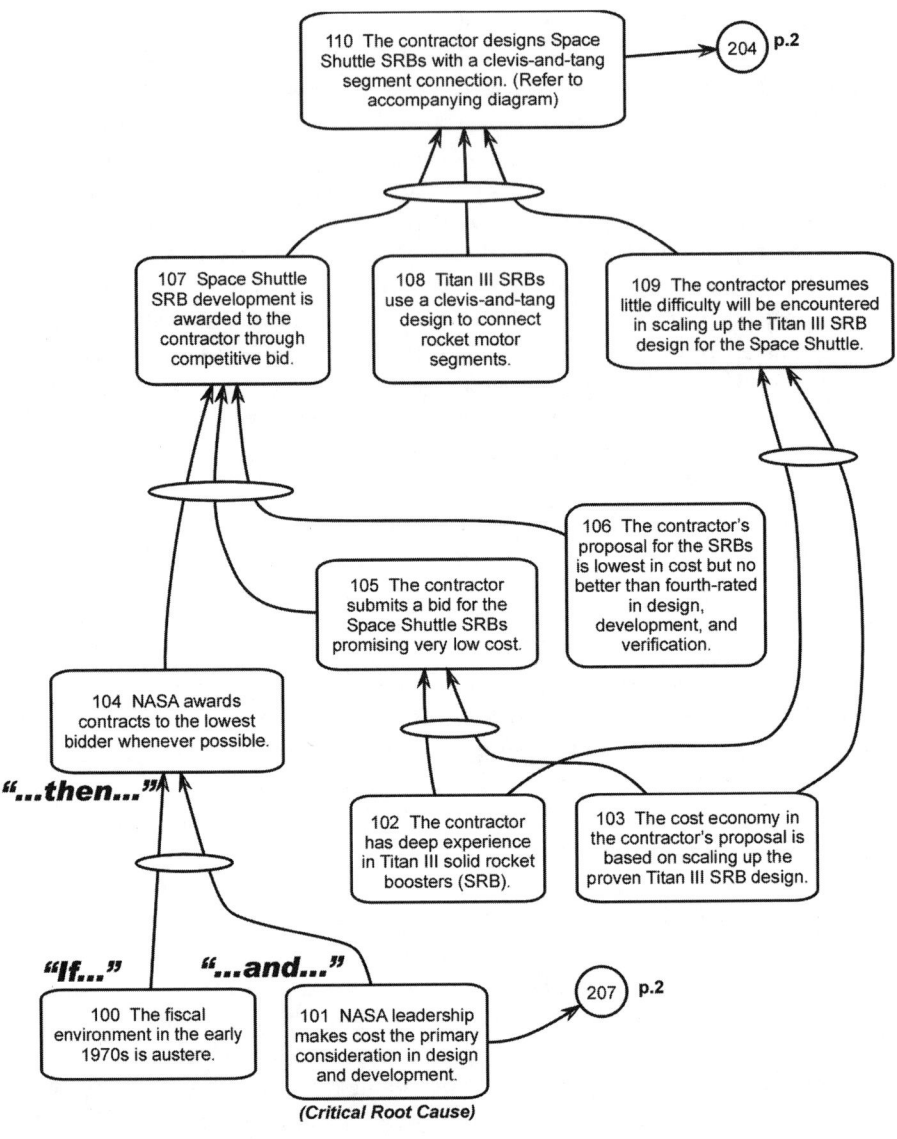

Figure 13.2a Current Reality Tree: The Challenger Accident. Modified from Dettmer, H.W. With permission.

reliable "workhorse" of space operations for many years. The contractor foresaw no difficulty in doing this.

But a major change occurred on the way to production of the Space Shuttle SRBs. The smaller Titan III booster had been assembled vertically. The larger Space Shuttle booster had to be assembled horizontally because existing frameworks were not large enough to accommodate the much taller Space Shuttle SRB. Laying the large-diameter thin-walled booster casing on its side caused the cylinder to flatten slightly, making it impossible to fit booster segments together at the joining point with the original design specifications.

The contractor's engineers immediately proposed redesigning the booster casing, but their proposal was rejected by both NASA and their own senior management because of the prohibitive cost and the schedule delay that would have been caused. The only other solution ("injection") they could think of at the time was to enlarge the receptacle space (clevis) in one of the booster segment joints to create a looser fit, allowing the "out of round" pieces to fit together. They did this.

Unfortunately, this solution produced a new problem. The increased space in the joints permitted horizontal assembly of the booster, but hydrostatic tests (high-pressure water) subsequently revealed major leaks around the booster's aft field joint at only half the pressure expected under actual launch conditions — an unacceptable disaster in the making. Because redesign of the booster had already been rejected, the contractor was forced to apply another "band-aid" to the already-compromised booster design: it was decided to "shim" the aft field joint to tighten the fit between segments. About 180 small wedges were inserted in the joint to aid in sealing it. As the world knows, this "injection" ultimately didn't work. But, in entities 204–205 (Figure 13.2b), we find the first place after contract award where the causality leading to the accident might have been broken with a combination of the conflict resolution diagram and TRIZ.

At each of several sequential events along the way, the contractor's engineers were faced with conflicts that could have been effectively expressed in a conflict resolution diagram. The first time they realized they had a problem was when they tried to fit two rocket motor segments together at the aft field joint. Because of the distortion of the booster casing's shape, the clevis and tang would not connect. This would not likely have been a problem with the smaller Titan III SRB, but the increase in size (cross-sectional area) of the larger shuttle SRBs coupled with horizontal assembly created the distortion. The CRD at this stage of development might have looked like Figure 13.3.

NASA and the contractor's senior management placed some restrictions on the engineers. They had to come up with some way to solve the problem without assembling the SRB vertically or redesigning it. This is not an unusual situation. In the real world, boundaries on potential solutions are often imposed with no room for negotiation.

The engineers decided to increase the specification for one part of the segment joint so that the out-of-round SRB segment would have

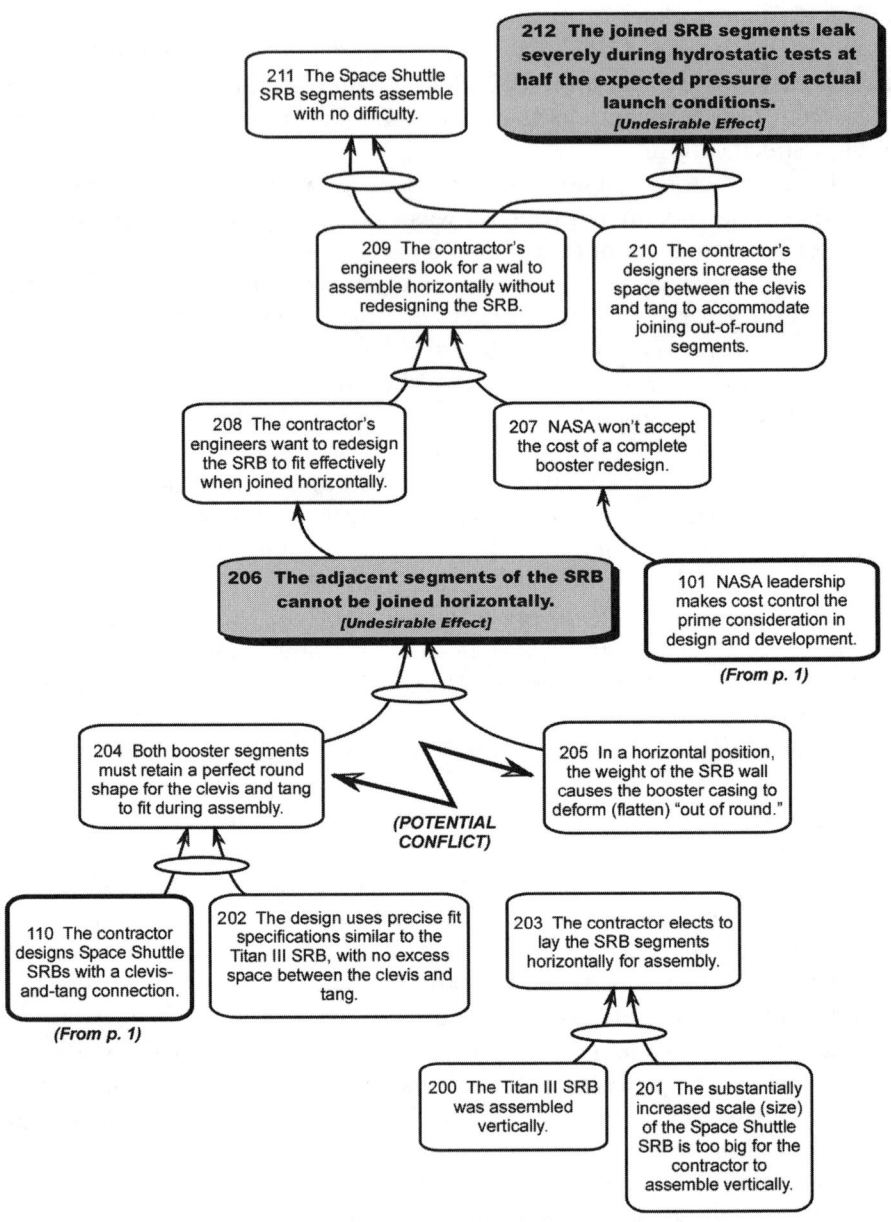

Figure 13.2b Current Reality Tree: The Challenger Accident. Modified from Dett-mer, H.W. With permission.

some "wiggle room" to fit the two halves together. This "injection" seemed to satisfy both requirements. But it created a new problem that

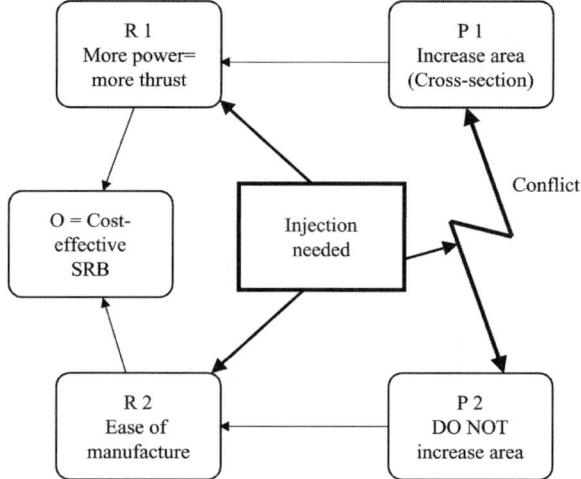

Assumptions:
1. More thrust requires larger boosters.
2. Larger boosters require larger cross-sections.
3. Ease of manufacture requires shape integrity.
4. Smaller cross-sections maintain a more circular shape.

Figure 13.3 Space Shuttle SRB Design Conflict, using the Conflict Resolution Diagram. Modified from Dettmer, H.W. With permission.

wasn't discovered until subsequent hydrostatic testing: the SRB now leaked at the modified joint and that leak posed an unacceptable flight hazard. In an attempt to salvage their first injection (enlarging the specifications), they decided to add another one: insert "shims" in the joint after the segments were mated to apply sufficient pressure to discourage the pressure leak. As we now know, this "band-aid" was an unsatisfactory solution — it created a safety problem later on. Let's see how TRIZ might have been applied to create a "breakthrough" idea that would have simultaneously satisfied the safety, cost and ease-of-assembly requirements.

The TRIZ ideal final result tool is used to keep focus on the broad scale problem. In this case, the ideal final result is that "the parts mate every time, simply, with no added processes and no leakage." Had the original team used a statement like this, they might have avoided the complex solutions that made the problem worse than the patch that atttempted to "fix the fix."

The CRD has identified contradictions present in the problem: shape (circularity) gets worse as area increases (improves). Another way of expressing this might be: "As area increases, manufacturability deteriorates."

Keeping costs as low as possible will be a decision rule for evaluating any potential solution. One of the oldest and simplest of the TRIZ tools, the 40 principles, can take us quickly to a family of creative solutions that resolve those contradictions, rather than compromising with them.

The technology of the time was such that increased power requirements (R1) demanded a larger booster, which translated to an increase in the cross-sectional area of the booster case (P1). This was a prerequisite imposed at the design stage by the laws of physics and chemistry. It left the engineers with only one option: figure out how to maintain the circular shape of the booster casing without sacrificing the cross-sectional area. So the two critical engineering parameters are area and shape: as the area of the cross-section improves, the shape of the cross-section deteriorates.

For this contradiction, the matrix suggests the use of principles 24 and 34.

- Principle 24: Intermediary. Use an intermediary carrier article or intermediary process. Merge one object temporarily with another (which can be easily removed).
- Principle 34: Discarding and recovering. Make portions of an object that have fulfilled their functions go away (discard by dissolving, evaporating, etc.) or modify these directly during operation. Conversely, restore consumable parts of an object directly in operation.

Combining these principles leads to the idea of forming the booster segments into a perfectly circular shape for mating by the use of a removable (Principle 34) mediator (Principle 24), or "jig" (see Figure 13.4). While the jig holds the circular shape, the segments are moved horizontally into position. The jig is then removed. The segments are successfully joined without having to relax the original fit specifications. The tight fit ensures seating and sealing of the O-rings with no pressure leakage and the Challenger explosion never occurs. Of course, it is much easier to do a case study like this with hindsight and this is intended as an example, not as a critique.

Those already familiar with the theory of constraints know that the conflict resolution diagram is particularly useful in resolving nontechnical conflicts, such as interpersonal organizational behavior or policy contentions.

The conflict resolution diagram in itself is a powerful tool for system improvement, as is TRIZ. Used together, each can reinforce the other to produce better, more creative solutions to complex conflict-related problems.

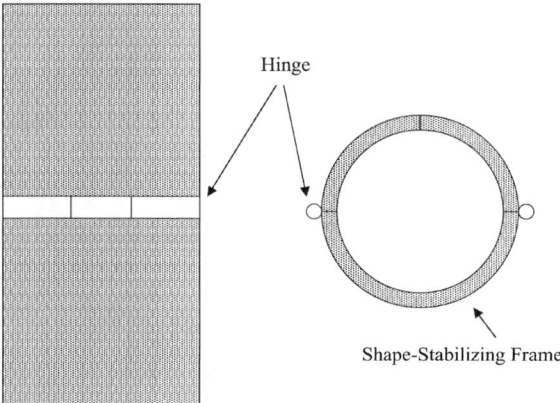

TRIZ Separation Principles 23 and 34
applied: Use a removable mediator

One Possible Solution:
HORIZONTAL ASSEMLY JIG

Figure 13.4   Using Principles 24 and 34 to find a way to maintain the shape of the booster during assembly. The parts of the booster would be inserted into this segmented cylindrical jig when open, then it would be closed, compressing the booster parts into shape so that they could be mated without leaking. The jig would then be removed and reused for the next assembly. Modified from Dettmer, H.W. With permission.

13.2   SUMMARY

- People who are familiar with the theory of constraints will recognize that the current reality tree and the conflict resolution diagram, either separately or together, show opportunities to use TRIZ to remove the cause of a problem.
- If TOC is already accepted in your organization, TRIZ can be introduced as a method used to make TOC even more effective.

REFERENCES

1. Goldratt, E.M. and Cox, J., *The Goal*, North River Press, Great Barrington, 1992.
2. Goldratt, E.M., *It's Not Luck*, North River Press, Great Barrington, 1994.
3. Goldratt, E.M., Schragenheim, E. and Ptak, C.A., *Necessary but not Sufficient*, North River Press, Great Barrington, 2000.
4. Dettmer, H.W., *Goldratt's Theory of Constraints*, ASQ Quality Press, 1996.

5. Dettmer, H.W., *Breaking the Constraints to World-Class Performance*, Milwaukee, ASQ Quality Press, 1998.
6. Domb, E. and Dettmer, H.W., Breakthrough innovation in conflict resolution, *The TRIZ Journal*, May 1999, www.triz-journal.com.
7. Moura, E.C., TOC trees help TRIZ, *The TRIZ Journal*, September 1999, www.triz-journal.com.

14

USING TRIZ WITH SIX SIGMA
AND OTHER QUALITY
IMPROVEMENT SYSTEMS

INTRODUCTION

The Six Sigma system for quality improvement in products, services, and processes is a business-based system of using statistical analysis and customer-focused methods. It has been demonstrated repeatedly that a company that moves from three sigma processes to six sigma processes increases its profitability by two to three orders of magnitude, and that companies that use the Design for Six Sigma (DFSS) process create products and services with much higher levels of customer satisfaction and technical quality than those that don't.[1-6] Coupling TRIZ with Six Sigma produces these powerful results faster, because the breakthrough problem-solving aspects of TRIZ can be focused on the profit opportunities identified by Six Sigma and the technology-forecasting aspects of TRIZ can be focused on planning new products at the right time in the product life cycle.

The breakthrough strategy of Six Sigma is different in vocabulary, but not in concept, from the *plan-do-check-act* method, usually known as PDCA, that has been used in quality improvement for the past 70 years and in human learning throughout our evolution.[7,8] The tools of TRIZ that are used in the improvement arena of Six Sigma are as shown in Table 14.1 as well as the relationship to the PDCA model.[1] The difference in emphasis between Six Sigma and conventional quality improvement methods is the focus at all levels of Six Sigma application on the business results of the proposed improvement.

Table 14.1 Opportunities to Apply TRIZ Occur in All of the Eight Phases of Six Sigma's Improvement Process (Sometimes Called "MAIC" after Phases C, D, E and F.)

Plan-Do-Check-Act Phase	Six Sigma Phase	TRIZ Opportunities
Plan	A. Recognize	Tool/object analysis, ideal final result
Plan	B. Define	Tool/object analysis
Plan	C. Measure	Develop measurement methods, improve instruments using technology forecasting and contradiction analysis
Plan	D. Analyze	Contradiction analysis
Plan-Do-Check	E. Improve	Create new product, process and service concepts (elimination of inherent or tradeoff contradictions, Scientific Effects)
Check-Act	F. Control	Same as C
Act	G. Standardize	Same as E, applied to service and product delivery system
Act, Plan	H. Integrate	Same as E, applied to whole system of improvement

Special vocabulary is used for the roles of the people involved in Six Sigma. Typically, Six Sigma *champions* identify improvement projects, and *black belts* lead project teams to conduct the analysis and improvement, using the eight steps identified as A–H in Table 14.1, or perform the activities themselves. *Green belts* are members of the project teams, or leaders of teams, or may occasionally perform projects themselves, if the full skills of a black belt are not required. *Master black belts* train the black belts and green belts and serve as their advisors as they conduct their projects. In many companies, the black belts and master black belts are now getting TRIZ training as well as classical Six Sigma training so they can accelerate the improvement process.

DFSS is used for either of two reasons:

1. To design new products, services or processes that can function at Six Sigma quality level, or at whatever quality level is selected for business reasons, using the Six Sigma criteria.
2. To improve existing products, services or processes, if the improvement requires a discontinuous redesign from the earlier system. Improvement beyond 4.5 sigma (the so-called "wall") often requires complete redesign.[5]

Table 14.2 The Relationship Between the TRIZ Tools and the Phases of Design for Six Sigma

DFSS Phase	TRIZ Tool
Multi-generational plan	Technology forecasting, tool/object analysis
Voice of customer and other elements of quality function deployment	Conflict resolution, ideal final result, development of measurement methods.
Concept development	All
Detailed design	All
Optimize	Conflict resolution, trimming, problem solving
Validate/implement	Same

Typically, DFSS is merged with the company's previous methods of product development initially, and a Six Sigma methodology is developed after the company has extensive experience with pilot projects. Table 14.2 lists the phases of DFSS and the TRIZ tools that are useful in each phase.

14.1 METHODS OF INTRODUCING TRIZ INTO SIX SIGMA

Six Sigma is a very highly structured system with a hierarchy of champions or project sponsors, master black belts, black belts, and green belts with defined levels of knowledge of business, identification of opportunities, statistical processes for analysis and control, and improvement at each level. Companies have inserted TRIZ into this process at a number of different points and in many different ways (training, workshop, consulting, etc.).

Motorola, the company that developed the Six Sigma process from its earlier quality improvement initiatives in the late 1980s, is at the least-structured end of the spectrum of methods of incorporating TRIZ into Six Sigma. TRIZ is taught and facilitated through the intellectual property organization. Six Sigma methods are taught and facilitated through a separate Six Sigma Organization. Black belts frequently study TRIZ and use TRIZ methods to solve their problems, but no joint curriculum exists.

General Electric and Allied Signal/Honeywell, the companies most famous for the economic benefits of their Six Sigma systems, have been similarly loosely structured. Many pockets of TRIZ knowledge exist within both companies, using a variety of TRIZ-derived methods and software systems, and applying TRIZ to Six Sigma projects.

The Ford Motor Company has used TRIZ methods in a variety of ways since the early 90s. They trained 400 people a year in the USIT (unified structured innovative thinking) version of TRIZ. In 2000–2001, Ford introduced TRIZ into the pilot project stage of their Six Sigma process.

Dow Chemical Company is also piloting the use of TRIZ in both DFSS and the MAIC (process improvement) system in 2001–2002. More than 150 research and development staff, including six master black belts and ten black belts received TRIZ training and applied it to their projects. For MAIC, a brief overview of TRIZ is presented to the black belts in their training, and they then decide whether to enroll in the TRIZ classes or to call on a TRIZ expert when their project team needs an innovative solution to a problem. For DFSS, the success of the pilot projects, which integrated QFD with TRIZ, led to the decision that all master black belts would be trained in TRIZ for problem solving and technology forecasting, as well as in the use of TRIZ-related software. They will then use TRIZ as needed, and teach TRIZ to the black belts and to members of the DFSS teams.

Delphi Automotive Systems is at the highly structured end of the spectrum of relationships between TRIZ and Six Sigma. TRIZ is used repeatedly in the design phase and in the process optimization phase. The overlap of tools and techniques between TRIZ, design of experiments, quality function deployment, and other methods in the identification of the ideal system, function analysis, and iterative improvement is emphasized throughout the Delphi training program and the use of the DFSS process.[9]

There are many other quality improvement initiatives in active use worldwide. Total quality management (TQM) evolved in the 1980s, was in wide use in the early 90s, and emphasized the need for quality in the *total* business—planning, management, sales, service, employee relationships, product development, etc. — as well as in the production area that was usually the focus of quality efforts. TQM is now used widely in healthcare and education quality initiatives, as well as in business. ISO-9000 and the related standards QS-9000 and AS-9000 initially emphasized the need for standardization and documentation, but in their revisions in 1999 and 2000, they placed much more emphasis on understanding customer needs and on continuous improvement based on customer and technical data. Many companies and government agencies have quality initiatives without formal names — they have been committed to customer-focused, business-focused quality improvement for so long that it has become a part of the organization's culture, not a separate "quality thing."[8] TRIZ is helpful in all these processes. The obvious way to use TRIZ is to fix technical problems with products and services. Less obvious ways include the following:

- Use TRIZ to find a creative way to get the customers' input. A classic problem for small companies doing international business is having no budget to travel to the customers' locations, to listen to and observe the customers. A problem for all companies in international business is lack of knowledge of their customers' language and culture. One Scandinavian electronics company found a TRIZ solution (using a *resource*) to the QFD challenge of listening to their female Japanese customers. They trained female Japanese employees of a subsidiary, who had done only production work, in customer interview skills. The results were excellent.[10]
- Use TRIZ to resolve the conflicts between the customers' needs and the organization's traditional way of doing things. Many electronic business ideas for customer direct access to consulting firm databases are emerging directly from this research.[11]

If your organization has a successful structured improvement process such as TOC or Six Sigma or any of the other improvement systems, the best way to introduce TRIZ may be as a family of tools that can help you resolve conflicts creatively. Once the organization has learned to appreciate TRIZ for problem solving, the expanded use of TRIZ for technology forecasting and strategic planning will be natural.

If your organization has no structured improvement process, the best way to introduce TRIZ is usually in each functional area as a problem-solving tool for that function. Once TRIZ is well established in certain key functions, it will spread to the rest of the company because it has proven its validity. Where to begin is a challenge that is specific to each company — we have seen successful implementations that started in manufacturing, engineering, customer service, sales, service development, warranty service, quality control, knowledge management, and intellectual property management, among others.

REFERENCES

1. Domb E., The Role of TRIZ in Six Sigma Management, TRIZCON2000, The Altshuller Institute, May, 2000.
2. Harry, M. and Schroeder, R., *Six Sigma*, Doubleday, New York, 2000.
3. Fisher, A., Rules for Joining the Cult of Perfectability, *Fortune*, Feb. 7, 2000.
4. Harry, M., A new definition aims to connect quality with financial performance, *Quality Progress*, January, 2000.
5. Perez-Wilson, M., *Six Sigma,* Scottsdale AZ, Advanced Systems Consultants, 1999.
6. Pande, P.S., Neuman, R.P. and Cavanagh, R.R., *The Six Sigma Way*, McGraw-Hill, New York, 2000.
7. Mann, D., Contradiction Chains, *The TRIZ Journal,* January, 2000, www.triz-journal.com.
8. Domb, E., Increase Creativity to Improve Quality: TRIZ and the Baldrige Award Criteria, TRIZCON1999, The Altshuller Institute, March, 1999.
9. Brown, A. Jr., The Role of Robust Engineering in Innovation and Continuous Improvement Methodologies, Presented at the ASI Six Sigma Symposium, October, 2000.
10. 11th Annual Symposium of the Quality Function Deployment Institute, June, 2000.
11. Domb, E. and Mann, D., Using TRIZ to Overcome Business Contradictions: Profitable E-Commerce, *The TRIZ Journal*, April, 2001, www.triz-journal.com, and *Proc. Portland Int. Conf. on Managing Engrg. Tech.*, July 2001.

15

BOOK SUMMARY: CREATIVE PROBLEM SOLVING IN A NUTSHELL

A compact graphical model for problem solving was introduced in Chapter 2 and repeated many times. The model helps you learn and understand TRIZ so you can use the methodology in situations when you don't have the book in hand. To remember the essential steps, the following list may help:

1. Model the system and the problem. Don't try to jump directly to the solution. Remember that you can download blank worksheets for modeling the problem and following the whole agenda for problem solving at http://www.triz-journal.com/SimplifiedTRIZ/.
2. Seek the contradictions behind the problem. Particularly, try to find one primary inherent contradiction. The contradiction should not be hidden or weakened.
3. Map the resources of the system. Try to find an invisible reserve. Don't be satisfied by easy solutions that require making the system more complex.
4. Formulate the ideal final result. Don't be satisfied by a conventional compromise.
5. Check the solution against the criteria of ideality. Ask: "What is the primary contradiction? Is it solved?"
6. Develop the solution further and improve it. It is cheaper and easier to repeat the solution cycle many times than to try to implement an incomplete idea.

7. Use the patterns of evolution and principles for innovation to improve solutions, and as independent tools to generate new ideas.
8. Integrate TRIZ with other methods already in use in the organization, such as QFD, TOC, Six Sigma, etc.
9. Look at TRIZ as an evolving theory, not as a rigid formula. Critically evaluate the TRIZ theory and methods. Improve your tools continuously — and publish what you've done, so others can benefit.

Summarizing the steps of the use and implementation we would like also recall the background of TRIZ, considered in the beginning of the book. If solutions that give many benefits and cost nothing are suggested, the immediate reaction may be that this is yet another hype. Try a simple exercise: recall your own personal bright moments in your career or business. You surely have examples of very good solutions. They may be good deals when the customer has been especially happy and you have made a big profit. Or perhaps you have managed to find a win–win solution to some difficult conflict between people so that everybody was satisfied. Maybe you remember a good engineering solution when the numbers of parts and operations were drastically decreased, performance improved, and costs cut. Any solution to some problem that was important to you will serve as an example.

Now think about what made your best solutions so different from everyday answers. Undoubtedly, your best solutions have given very big benefits compared with costs and possible harmful side effects. They have increased the ideality of the system, which improves when the increase in benefits outweighs any increase in costs of the system and the harm done by the system. We have learned that there are solutions near ideality, and everybody has sometimes achieved them. Use your own examples to teach these concepts to others.

In the first chapter, we stressed that the point of this book is to learn how to create and recognize good solutions. Analyzing the common features of good solutions across industries, we can find tools for developing new ideas. We compared the tools with vehicles. A mediocre driver moves faster than the best runner. Other studies of creativity put emphasis on the characteristics of people who are good problem solvers.

Discussions of creativity and innovation are often rather fruitless, because only the improvement of people is considered. In Chapter 1, we referred to theories X and Y, presented several decades ago by McGregor. The assumption X says that, "People must be coerced, controlled, directed, threatened with punishment to get them to put forth adequate effort."[1] This assumption flourishes in stories of creative effort. You have probably many times heard the testimonials of handicapped people who say that the accident that injured them helped them to accomplish some incredible feat. Popular as these statements may be, it is easy to disprove them. There are true stories of handicapped people who "turned lemons into lemonade," but there are also many stories of people who suffered without producing any extraordinary achievements. People get results *in spite of* accidents, persecution or poverty, not *because of* hardships and disasters.

Equally popular is the belief that the main obstacle blocking innovation is the lack of resources, and that pouring money into research and design will increase the output of "innovation." Academic studies, such as several by Michael Porter, and the most recent by Porter and Stern,[2] reinforce the belief that a rich environment, with capital, suppliers of components, and an innovation-supporting infrastructure are necessary. Certainly this assumption has more appeal to common sense than the idea that accidents or poverty stimulate creativity. But something more is needed.

Another comparison illustrates this. Literacy is a mental tool, widespread in developed countries. It is obvious that, in a poor country with huge illiterate population, one of the first things needed for economic development is to give workers literacy — the mental vehicle. Money will not induce people to be more creative — they already have been very creative to survive and feed their families in poverty.

In a complex, highly developed society, the needs are less obvious, but a similar picture can be seen. Trying to improve creativity by giving people more time and money for simple brainstorming is like trying to turn poor runners into marathoners by giving them time to practice and offering money for winning races.

While recognizing that people differ in their natural ability to be creative, in this book we focus on the tools that everyone can use, regardless of natural ability. You have learned to use a new tool, the law of increasing ideality and the definition of the ideal final result.

REFERENCES

1. McGregor, D.; *The Human Side of Enterprise*, McGraw-Hill, New York, 1960, 34.
2. Porter, M.A. and Stern, S., Innovation: Location Matters. *MIT Sloan Management Review*, Summer, 2001.

16

GET STARTED

TRIZ works. If you have done the exercises as you read this book, you know that TRIZ will help you find innovative solutions to problems, help you understand the evolution of systems, and help you develop more ideas faster.

Many other tools of TRIZ were not included in this book because our goal is to help you start using TRIZ quickly. Once you have mastered the TRIZ methods presented here by applying them to real problems in your business and your personal life, you may want to learn more about advanced tools of TRIZ. But our advice is not to try to learn more tools now. As the title of this chapter says, get started.

We invite our readers to send questions and comments to us at http://www.triz-journal.com/simplified TRIZ. Your comments will help us improve future editions of this book. But you won't have any stories to tell or questions to ask if you don't *GET STARTED*.

GLOSSARY

One of the purposes of this book is to keep terminology as simple and exact as possible. New terms have replaced some older TRIZ words. For example, we speak of "tradeoff" instead of "technical contradiction". The first criterion for selecting terms is that they reflect the subject matter adequately and are compatible with everyday language, and with professional language in industry.

Compatibility with old TRIZ terms has taken second place. Readers interested in the older terminology used in the TRIZ community should consult the glossaries prepared by Fey[1] and Savransky.[2]

Action The influence of one component on another, particularly the influence of the tool on the object.

ARIZ Acronym for the Russian words *"algorithm rezhenija izobretatelskih zadach."* An English translation is ASIP (algorithm for solving inventive problems). A step-by-step guide was developed by Altshuller for the analysis and resolving of contradictions. Altshuller and his team developed several versions of ARIZ 1956–1985.

ASIP See **ARIZ**.

Auxiliary resource Resource that changes the **principal resources** so that the **inherent contradiction** is resolved.

Conflict See **Contradiction**.

Contradiction Opposition between things or properties of things. There are two kinds of contradictions: (1) **Tradeoff**, a situation in which if something good happens, something bad also happens. Or, if something good gets better, something undesirable gets worse. (2) **Inherent contradiction** is a situation in which one thing has two opposite properties.

Convolution Decreasing the number of parts and operations in the system so that useful features and functions are retained. See **Expansion**.

Engineering contradiction See **Tradeoff**.

Expansion Increasing the number of parts and operations in the system so that useful features and functions are increased.

Feature A property of the system.

Field See **Interaction**.

Function The term function is a diffused concept with many meanings: (1) Interaction including the **Action** and the **Object** of action (the motorcycle moves the person); (2) the purpose of the action (the purpose of the motorcycle may be to entertain the person); and (3) the result of the action (the motorcycle generates or produces noise and exhaust gas).

Ideal final result (1) The solution that removes the **Contradiction** using the **Resources** in the system and its environment; (2) a description of the desired outcome, without use of jargon, that emphasizes achievement of the benefits of the system; (3) Algebraically, the situation for the ideality equation when the denominator approaches zero: Ideality = Σ Benefits/(Σ Cost + Σ Harm).

Inherent contradiction A situation in which one thing has two opposite properties. See also **Contradiction** and **Tradeoff**.

Interaction Influence of the components of a system on each other. See **Action**.

Instrument See **Tool**.

Model An idealized concise description of phenomena and problems. A model contains relevant parts, and connections between them and explains the evolution of the system.

Object The component of a system that is influenced or acted on by the **Tool**.

Pattern of evolution A regularity discovered in the evolution of the system. Repetition of a sequence of similar events in the history of a system.

Physical contradiction See **Inherent contradiction**.

Principal resource Most important **resource** containing **inherent contradiction**. See **Resources**.

Principle Principle for innovation. A generic solution applicable in many industries. The most widely used set of these solutions is the list of the 40 principles.

Psychological inertia The resistance to thinking a new way. By analogy to physical inertia; thoughts continue in the same pattern unless disrupted by a force.

Resources Things, information, energy, time, space, or properties of the materials that are already in or near the environment of the system, and are available for the resolution of the **contradiction** and achieving the **ideal final result**.

Standard solution Typical transformation of the system, improving it and removing the **contradiction**. Altshuller and his team developed the 76 standard solutions.

Substance and field resources See **Resources**.

System The set of objects and the interactions between them, having features or properties not reduceable to the features or properties of separate objects. The system of objects is more than the sum of these objects, because of the interactions between them. The set of interacting **tools** and **objects**.

Technical contradiction See **Tradeoff**.

TIPS See **TRIZ**.

Tool Component that influences or acts on the **object**.

Tradeoff If something good happens, something bad also happens. Or, if something good gets better, something undesirable gets worse.

Trimming See **Convolution**.

TRIZ Theory of inventive problem solving (TIPS). Acronym for the Russian words "*teorija rezhenija izobretatelskih zadach.*"

Zone of Proximal Development Solutions that are possible but have not yet been developed.

REFERENCES

1. Fey, V., TRIZ Glossary, *Izobretenia*, II, 15, 200.
2. Savransky, S.D., *Engineering of Creativity*, CRC, Boca Raton, 2000.

INDEX